BASIC DATA ANALYSIS FOR TIME SERIES WITH R

BASIC DATA ANALYSIS FOR TIME SERIES WITH R

DEWAYNE R. DERRYBERRY

Department of Mathematics and Statistics
Idaho State University
Boise, ID

For general information on our other products and services or for technical support, please contact our Customer Care Department within the United States at (800) 762-2974, outside the United States at (317) 572-3993 or fax (317) 572-4002.

Wiley also publishes its books in a variety of electronic formats. Some content that appears in print may not be available in electronic formats. For more information about Wiley products, visit our web site at www.wiley.com.

Library of Congress Cataloging-in-Publication Data:

Derryberry, DeWayne R., author.
    Basic data analysis for time series with R / DeWayne R. Derryberry, Department of Mathematics and Statistics, Idaho State University, Voise, ID.
        pages cm
    Includes bibliographical references and index.
    ISBN 978-1-118-42254-0 (hardback)
    1. Time-series analysis–Data processing.   2. R (Computer program language)   I. Title.
    QA280.D475 2014
    001.4′2202855133–dc23

                                                                                        2014007300

Printed in the United States of America

ISBN: 9781118422540

10 9 8 7 6 5 4 3 2 1

CONTENTS

PREFACE   xv

ACKNOWLEDGMENTS   xvii

PART I   BASIC CORRELATION STRUCTURES

1   R Basics   3

     1.1   Getting Started, 3
     1.2   Special R Conventions, 5
     1.3   Common Structures, 5
     1.4   Common Functions, 6
     1.5   Time Series Functions, 6
     1.6   Importing Data, 7
     Exercises, 7

2   Review of Regression and More About R   8

     2.1   Goals of this Chapter, 8
     2.2   The Simple(ST) Regression Model, 8
            2.2.1   Ordinary Least Squares, 8
            2.2.2   Properties of OLS Estimates, 9
            2.2.3   Matrix Representation of the Problem, 9
     2.3   Simulating the Data from a Model and Estimating the Model
            Parameters in R, 9
            2.3.1   Simulating Data, 9
            2.3.2   Estimating the Model Parameters in R, 9

2.4   Basic Inference for the Model, 12

2.5   Residuals Analysis—What Can Go Wrong…, 13

2.6   Matrix Manipulation in R, 15

    2.6.1   Introduction, 15

    2.6.2   OLS the Hard Way, 15

    2.6.3   Some Other Matrix Commands, 16

Exercises, 16

**3   The Modeling Approach Taken in this Book and Some Examples
of Typical Serially Correlated Data**   **18**

3.1   Signal and Noise, 18

3.2   Time Series Data, 19

3.3   Simple Regression in the Framework, 20

3.4   Real Data and Simulated Data, 20

3.5   The Diversity of Time Series Data, 21

3.6   Getting Data Into R, 24

    3.6.1   Overview, 24

    3.6.2   The Diskette and the *scan()* and *ts()* Functions—New
York City Temperatures, 25

    3.6.3   The Diskette and the *read.table()* Function—The
Semmelweis Data, 25

    3.6.4   Cut and Paste Data to a Text Editor, 26

Exercises, 26

4   Some Comments on Assumptions   **28**

4.1   Introduction, 28

4.2   The Normality Assumption, 29

    4.2.1   Right Skew, 30

    4.2.2   Left Skew, 30

    4.2.3   Heavy Tails, 30

4.3   Equal Variance, 31

    4.3.1   Two-Sample *t*-Test, 31

    4.3.2   Regression, 31

4.4   Independence, 31

4.5   Power of Logarithmic Transformations Illustrated, 32

4.6   Summary, 34

Exercises, 34

5   The Autocorrelation Function And AR(1), AR(2) Models   **35**

5.1   Standard Models—What are the Alternatives to *White Noise*?, 35

5.2   Autocovariance and Autocorrelation, 36

    5.2.1   Stationarity, 36

    5.2.2   A Note About Conditions, 36

5.2.3   Properties of Autocovariance, 36

5.2.4   White Noise, 37

5.2.5   Estimation of the Autocovariance and Autocorrelation, 37

5.3   The *acf( )* Function in R, 37

5.3.1   Background, 37

5.3.2   The Basic Code for Estimating the Autocovariance, 38

5.4   The First Alternative to White Noise: Autoregressive Errors—AR(1), AR(2), 40

5.4.1   Definition of the AR(1) and AR(2) Models, 40

5.4.2   Some Preliminary Facts, 40

5.4.3   The AR(1) Model Autocorrelation and Autocovariance, 41

5.4.4   Using Correlation and Scatterplots to Illustrate the AR(1) Model, 41

5.4.5   The AR(2) Model Autocorrelation and Autocovariance, 41

5.4.6   Simulating Data for AR(m) Models, 42

5.4.7   Examples of Stable and Unstable AR(1) Models, 44

5.4.8   Examples of Stable and Unstable AR(2) Models, 46

Exercises, 49

6   The Moving Average Models MA(1) And MA(2)   **51**

6.1   The Moving Average Model, 51

6.2   The Autocorrelation for MA(1) Models, 51

6.3   A Duality Between MA(l) And AR(m) Models, 52

6.4   The Autocorrelation for MA(2) Models, 52

6.5   Simulated Examples of the MA(1) Model, 52

6.6   Simulated Examples of the MA(2) Model, 54

6.7   AR(m) and MA(l) model *acf( )* Plots, 54

Exercises, 57

PART II   ANALYSIS OF PERIODIC DATA AND MODEL SELECTION

7   Review of Transcendental Functions and Complex Numbers   **61**

7.1   Background, 61

7.2   Complex Arithmetic, 62

7.2.1   The Number i, 62

7.2.2   Complex Conjugates, 62

7.2.3   The Magnitude of a Complex Number, 62

7.3   Some Important Series, 63

7.3.1   The Geometric and Some Transcendental Series, 63

7.3.2   A Rationale for Euler's Formula, 63

7.4   Useful Facts About Periodic Transcendental Functions, 64

Exercises, 64

8   The Power Spectrum and the Periodogram   **65**

8.1   Introduction, 65
8.2   A Definition and a Simplified Form for $p(f)$, 66
8.3   Inverting $p(f)$ to Recover the C_k Values, 66
8.4   The Power Spectrum for Some Familiar Models, 68
    8.4.1   White Noise, 68
    8.4.2   The Spectrum for AR(1) Models, 68
    8.4.3   The Spectrum for AR(2) Models, 70
8.5   The Periodogram, a Closer Look, 72
    8.5.1   Why is the Periodogram Useful?, 72
    8.5.2   Some Naïve Code for a Periodogram, 72
    8.5.3   An Example—The Sunspot Data, 74
8.6   The Function *spec.pgram( )* in R, 75
Exercises, 77

**9   Smoothers, The Bias-Variance Tradeoff, and the Smoothed
Periodogram**   **79**

9.1   Why is Smoothing Required?, 79
9.2   Smoothing, Bias, and Variance, 79
9.3   Smoothers Used in R, 80
    9.3.1   The R Function *lowess( )*, 81
    9.3.2   The R Function *smooth.spline( )*, 82
    9.3.3   Kernel Smoothers in *spec.pgram( )*, 83
9.4   Smoothing the Periodogram for a Series With a Known and
Unknown Period, 85
    9.4.1   Period Known, 85
    9.4.2   Period Unknown, 86
9.5   Summary, 87
Exercises, 87

10   A Regression Model for Periodic Data   **89**

10.1   The Model, 89
10.2   An Example: The NYC Temperature Data, 91
    10.2.1   Fitting a Periodic Function, 91
    10.2.2   An Outlier, 92
    10.2.3   Refitting the Model with the Outlier Corrected, 92
10.3   Complications 1: CO_2 Data, 93
10.4   Complications 2: Sunspot Numbers, 94
10.5   Complications 3: Accidental Deaths, 96
10.6   Summary, 96
Exercises, 96

11   Model Selection and Cross-Validation   **98**

11.1 Background, 98
11.2 Hypothesis Tests in Simple Regression, 99
11.3 A More General Setting for Likelihood Ratio Tests, 101
11.4 A Subtlety Different Situation, 104
11.5 Information Criteria, 106
11.6 Cross-validation (Data Splitting): NYC Temperatures, 108
        11.6.1 Explained Variation, R^2, 108
        11.6.2 Data Splitting, 108
        11.6.3 Leave-One-Out Cross-Validation, 110
        11.6.4 AIC as Leave-One-Out Cross-Validation, 112
11.7 Summary, 112
Exercises, 113

12   Fitting Fourier series   **115**

12.1 Introduction: More Complex Periodic Models, 115
12.2 More Complex Periodic Behavior: Accidental Deaths, 116
        12.2.1 Fourier Series Structure, 116
        12.2.2 R Code for Fitting Large Fourier Series, 116
        12.2.3 Model Selection with AIC, 117
        12.2.4 Model Selection with Likelihood Ratio Tests, 118
        12.2.5 Data Splitting, 119
        12.2.6 Accidental Deaths—Some Comment on Periodic Data, 120
12.3 The Boise River Flow data, 121
        12.3.1 The Data, 121
        12.3.2 Model Selection with AIC, 122
        12.3.3 Data Splitting, 123
        12.3.4 The Residuals, 123
12.4 Where Do We Go from Here?, 124
Exercises, 124

13   Adjusting for AR(1) Correlation in Complex Models   **125**

13.1 Introduction, 125
13.2 The Two-Sample t-Test—UNCUT and Patch-Cut Forest, 125
        13.2.1 The Sleuth Data and the Question of Interest, 125
        13.2.2 A Simple Adjustment for t-Tests When the Residuals Are
                AR(1), 128
        13.2.3 A Simulation Example, 129
        13.2.4 Analysis of the Sleuth Data, 131
13.3 The Second Sleuth Case—Global Warming, A Simple Regression, 132
        13.3.1 The Data and the Question, 132
        13.3.2 Filtering to Produce (Quasi-)Independent Observations, 133

13.3.3 Simulated Example—Regression, 134

13.3.4 Analysis of the Regression Case, 135

13.3.5 The Filtering Approach for the Logging Case, 136

13.3.6 A Few Comments on Filtering, 137

13.4 The Semmelweis Intervention, 138

13.4.1 The Data, 138

13.4.2 Why Serial Correlation?, 139

13.4.3 How This Data Differs from the Patch/Uncut Case, 139

13.4.4 Filtered Analysis, 140

13.4.5 Transformations and Inference, 142

13.5 The NYC Temperatures (Adjusted), 142

13.5.1 The Data and Prediction Intervals, 142

13.5.2 The AR(1) Prediction Model, 144

13.5.3 A Simulation to Evaluate These Formulas, 144

13.5.4 Application to NYC Data, 146

13.6 The Boise River Flow Data: Model Selection With Filtering, 147

13.6.1 The Revised Model Selection Problem, 147

13.6.2 Comments on R^2 and R^2_{pred}, 147

13.6.3 Model Selection After Filtering with a Matrix, 148

13.7 Implications of AR(1) Adjustments and the "Skip" Method, 151

13.7.1 Adjustments for AR(1) Autocorrelation, 151

13.7.2 Impact of Serial Correlation on p-Values, 152

13.7.3 The "skip" Method, 152

13.8 Summary, 152

Exercises, 153

PART III   COMPLEX TEMPORAL STRUCTURES

14   The Backshift Operator, the Impulse Response Function, and General ARMA Models   159

14.1 The General ARMA Model, 159

14.1.1 The Mathematical Formulation, 159

14.1.2 The *arima.sim( )* Function in R Revisited, 159

14.1.3 Examples of ARMA(m,l) Models, 160

14.2 The Backshift (Shift, Lag) Operator, 161

14.2.1 Definition of B, 161

14.2.2 The Stationary Conditions for a General AR(m) Model, 161

14.2.3 ARMA(m,l) Models and the Backshift Operator, 162

14.2.4 More Examples of ARMA(m,l) Models, 162

14.3 The Impulse Response Operator—Intuition, 164

14.4 Impulse Response Operator, $g(B)$—Computation, 165

14.4.1 Definition of $g(B)$, 165

14.4.2 Computing the Coefficients, g_j., 165

14.4.3 Plotting an Impulse Response Function, 166

14.5 Interpretation and Utility of the Impulse Response Function, 167
Exercises, 167

**15 The Yule–Walker Equations and the Partial Autocorrelation
Function**   **169**

15.1 Background, 169
15.2 Autocovariance of an ARMA(m,l) Model, 169
    15.2.1 A Preliminary Result, 169
    15.2.2 The Autocovariance Function for ARMA(m,l) Models, 170
15.3 AR(m) and the Yule–Walker Equations, 170
    15.3.1 The Equations, 170
    15.3.2 The R Function *ar.yw()* with an AR(3) Example, 171
    15.3.3 Information Criteria-Based Model Selection Using *ar.yw()*, 173
15.4 The Partial Autocorrelation Plot, 174
    15.4.1 A Sequence of Hypothesis Tests, 174
    15.4.2 The *pacf()* Function—Hypothesis Tests Presented in a Plot, 174
15.5 The Spectrum For Arma Processes, 175
15.6 Summary, 177
Exercises, 178

16 Modeling Philosophy and Complete Examples   **180**

16.1 Modeling Overview, 180
    16.1.1 The Algorithm, 180
    16.1.2 The Underlying Assumption, 180
    16.1.3 An Example Using an AR(m) Filter to Model MA(3), 181
    16.1.4 Generalizing the "Skip" Method, 184
16.2 A Complex Periodic Model—Monthly River Flows, Furnas
    1931–1978, 185
    16.2.1 The Data, 185
    16.2.2 A Saturated Model, 186
    16.2.3 Building an AR(m) Filtering Matrix, 187
    16.2.4 Model Selection, 189
    16.2.5 Predictions and Prediction Intervals for an AR(3) Model, 190
    16.2.6 Data Splitting, 191
    16.2.7 Model Selection Based on a Validation Set, 192
16.3 A Modeling Example—Trend and Periodicity: CO_2 Levels at
    Mauna Lau, 193
    16.3.1 The Saturated Model and Filter, 193
    16.3.2 Model Selection, 194
    16.3.3 How Well Does the Model Fit the Data?, 197
16.4 Modeling Periodicity with a Possible Intervention—Two Examples, 198
    16.4.1 The General Structure, 198
    16.4.2 Directory Assistance, 199
    16.4.3 Ozone Levels in Los Angeles, 202

16.5 Periodic Models: Monthly, Weekly, and Daily Averages, 205

16.6 Summary, 207

Exercises, 207

PART IV   SOME DETAILED AND COMPLETE EXAMPLES

17   Wolf's Sunspot Number Data   **213**

17.1 Background, 213

17.2 Unknown Period ⇒ Nonlinear Model, 214

17.3 The Function *nls( )* in R, 214

17.4 Determining the Period, 216

17.5 Instability in the Mean, Amplitude, and Period, 217

17.6 Data Splitting for Prediction, 220

17.6.1 The Approach, 220

17.6.2 Step 1—Fitting One Step Ahead, 222

17.6.3 The AR Correction, 222

17.6.4 Putting it All Together, 223

17.6.5 Model Selection, 223

17.6.6 Predictions Two Steps Ahead, 224

17.7 Summary, 226

Exercises, 226

18   An Analysis of Some Prostate and Breast Cancer Data   **228**

18.1 Background, 228

18.2 The First Data Set, 229

18.3 The Second Data Set, 232

18.3.1 Background and Questions, 232

18.3.2 Outline of the Statistical Analysis, 233

18.3.3 Looking at the Data, 233

18.3.4 Examining the Residuals for AR(m) Structure, 235

18.3.5 Regression Analysis with Filtered Data, 238

Exercises, 243

19   Christopher Tennant/Ben Crosby Watershed Data   **245**

19.1 Background and Question, 245

19.2 Looking at the Data and Fitting Fourier Series, 246

19.2.1 The Structure of the Data, 246

19.2.2 Fourier Series Fits to the Data, 246

19.2.3 Connecting Patterns in Data to Physical Processes, 246

19.3 Averaging Data, 248

19.4 Results, 250

Exercises, 250

20   Vostok Ice Core Data   **251**

20.1 Source of the Data, 251
20.2 Background, 252
20.3 Alignment, 253
    20.3.1 Need for Alignment, and Possible Issues Resulting from
           Alignment, 253
    20.3.2 Is the Pattern in the Temperature Data Maintained?, 254
    20.3.3 Are the Dates Closely Matched?, 254
    20.3.4 Are the Times Equally Spaced?, 255
20.4 A Naïve Analysis, 256
    20.4.1 A Saturated Model, 256
    20.4.2 Model Selection, 258
    20.4.3 The Association Between CO_2 and Temperature Change, 258
20.5 A Related Simulation, 259
    20.5.1 The Model and the Question of Interest, 259
    20.5.2 Simulation Code in R, 260
    20.5.3 A Model Using all of the Simulated Data, 261
    20.5.4 A Model Using a Sample of 283 from the Simulated Data, 262
20.6 An AR(1) Model for Irregular Spacing, 265
    20.6.1 Motivation, 265
    20.6.2 Method, 266
    20.6.3 Results, 266
    20.6.4 Sensitivity Analysis, 267
    20.6.5 A Final Analysis, Well Not Quite, 268
20.7 Summary, 269
Exercises, 270

Appendix A   Using Datamarket   **273**

A.1 Overview, 273
A.2 Loading a Time Series in Datamarket, 277
A.3 Respecting Datamarket Licensing Agreements, 280

Appendix B   AIC is PRESS!   **281**

B.1 Introduction, 281
B.2 PRESS, 281
B.3 Connection to Akaike's Result, 282
B.4 Normalization and R^2, 282
B.5 An example, 283
B.6 Conclusion and Further Comments, 283

Appendix C   A 15-Minute Tutorial on Nonlinear Optimization   **284**

C.1 Introduction, 284
C.2 Newton's Method for One-Dimensional Nonlinear Optimization, 284

C.3   A Sequence of Directions, Step Sizes, and a Stopping Rule, 285

C.4   What Could Go Wrong?, 285

C.5   Generalizing the Optimization Problem, 286

C.6   What Could Go Wrong—Revisited, 286

C.7   What Can be Done?, 287

REFERENCES   **291**

INDEX   **293**

PREFACE

WHAT THIS BOOK IS ABOUT

- Real, often very messy, data
- Using the popular R programming language
- Both data analysis and mathematical exercises
- Modern principles of model selection and model building, such as information criteria and data splitting, and hypothesis testing, as well as a complete approach to model selection, independent of the criteria
- Graphical displays of the data

MOTIVATION

A few years ago I was asked to teach a class on time series. The students in the class would be undergraduate statistics majors, graduate mathematics majors, and graduate students from a variety of science departments around the campus. My major concern in the course was that by the end of the course, the students be able to actually analyze data. In fact, I am of the belief that any applied statistics course, upon completion, in which the best students cannot independently analyze data, is a waste of the student's time and money.

Although this sounds like a reasonable goal, it is surprising how inadequate most time series books look, given this task. Most time series books fall into two categories: books about mathematical models with data sets used to illustrate a mathematical idea, and books on forecasting with recipes for making short-term predictions.

The only presentation close to what I wanted was Chapter 15 of *The Statistical Sleuth* (2002) by Ramsey and Schafer. However, this chapter is limited to what we will come to denote the AR(1) model and lacks the mathematical details to go further.

There is a second motivation for the book. Statisticians lament the misuse and poor use of statistical methods in nonstatistical journals. Of course, in this regard, as Mark Twain would say, "everyone complaints about it, but nobody does anything." In fact, most statisticians are too busy getting tenure and developing methods magnitudes of sophistication beyond what colleagues in nonstatistical journals are doing. Not many statisticians have made much effort toward translating the really basic ideas from decades ago into something nonstatisticians can use.

This book is a result of teaching a time series class twice. There are two ways this book is different from most other books on time series: (1) it is dated, there is nothing in this book that was not known by 1975, perhaps 1955 (in terms of mathematical and statistical theory) and (2) it is all about data. To the extent that this book is viewed as a generalization and extension of Chapter 15 of *The Statistical Sleuth*, it is a success.

REQUIRED BACKGROUND

The book presumes knowledge of linear algebra, data analysis, and basic computer programming. The book is aimed at two distinct audiences; as a textbook for a senior or slightly more advanced special topics course in statistics or as reference for statisticians who want to learn sufficient theory to analyze data with serial correlation, but have had little or no special training in time series. It is assumed that writing (a lot of) code is part of a modern applied statistics class at the senior level, and the student and instructor will write a lot of code if they use this book.

A COUPLE OF ODD FEATURES

In two cases (Wolf's sunspot numbers and Semmelweis), background material has been quoted from Wikipedia. In fact, the information cited is well known and citing Wikipedia requires no special permissions. Many of the data sets in the book come from simulations. Both graphs and numerical summaries are displayed. In such cases, the graphical display may have been revised several times in the course of writing the book. Rather than changing the numerical summaries each time, which would sometimes lead to extensive revisions, the author just left the first analysis and changed the graph with each new simulation. A gifted savant, as in the movie *Rainman*, might find some subtle mismatch between some graph of simulated data and the reported numerical values, but most of us would never detect any discrepancies.

ACKNOWLEDGMENTS

First and foremost, I must acknowledge the patient and gracious students who took my classes: Jeremy Farrell, Michael Chow, Megan Loveland, Anthony Lock-Smith, Gee Mellisa Pricilia, Yevgeniy Ptukhin, Christie Solomon (Spring 2011); Theo Barnhart, Leland Davis, Meghan Fisher, Noelle Guernsey, Michael Jacobson, Andrew Jensen, Christopher Tennant, Tyler Yazzie, Ted Owens, Garrett Castle, Anthony Wilson (Fall 2012). Students who take a class based only on notes have a difficult task. There is no table of contents, nor is there an index. Notes are full of errors, typos, and graphs with poor labeling. Page numbers are in constant flux. Sometimes, the students get to follow the instructor down dead ends (if properly done, this really is a good way to teach mathematics). On the other hand, for those who can handle a heavy cognitive load, there is something to be said for learning with the instructor instead of learning from the instructor. Garrett Castle also worked for me one school year helping get the book in the final form.

A number of colleagues have been helpful in different ways. My chair, Bob Fisher, has been very helpful in scheduling so that I could teach the course twice and get the book written. Greg Snow would not know me, but he taught at the 2.5-day workshop that I attended on R at Brigham Young University in the summer of 2013, and this was extremely helpful.

Certainly, at some level, this book would never have the great data sets without the Time Series Data Library compiled by Rob Hyndman, Professor of Statistics at Monash University, and now housed at DataMarket. People do not get tenure for compiling data sets, but there is no substitute for great data.

Finally, one must thank those who shaped his or her thinking. Given the unorthodox nature of this book, thanking mentors is a tricky business. If the book is a success, the authors will realize they thanked too few mentors and others who were influential;

if the book is a failure, they will wish they had not mentioned anyone. However, at a bare minimum, I must thank Lawrence Mayer, Arizona State University, who imparted an odd mixture of wisdom and applied statistical expertise. He was certainly the best teacher I ever had. I must thank Paul Murtaugh, Oregon State University, who patiently directed my doctoral research (in survival analysis). Finally, I must thank Cliff Pereira and N. Scott Urquhart who mentored the Oregon State University consulting seminar when I was there.

The Statistical Sleuth has been a great influence, if only to show that books do not need to be written the way most books are. So I should thank former teachers Fred Ramsey and Dan Schafer. Whether this book is headed in a good direction or a bad direction, it at least shares with *The Statistical Sleuth* that it is headed in a different direction.

PART I

BASIC CORRELATION STRUCTURES

1

R BASICS

1.1 GETTING STARTED

Programming, writing code, is an integral part of modern data analysis. This book assumes some experience with a programming language. R is simply a high level programming language with several functions that perform specialized statistical analysis. Any R programmer may also write their own functions, but the focus of this book is using functions already in R and writing some basic code using these functions.

In this book, R code is always shaded. Related R output that follows is always underlined or, if extensive, presented in a table. Comments in R code follow the pound sign, #. When R functions are discussed in the text, they will be *italicized*.

Some basic statistical analysis might involve inputting some values and performing some basic statistical analysis. Following the conventions already stated,

```
# this is a comment
x <- c(0.1,1.1,2.3,4.1, 5.6, 8.3)  # create a vector of 6 values, these were just made up
y <- c(8.7, 6.5, 3.1, 3.3, −1.1, −0.9) #create a second vector
plot(x,y,xlab = "this places a label on the x axis", ylab = "this labels the y axis")
title("this is a demo")
# either "<-" or "=" can be used for assignments, but "<-" makes more sense.
# The reader should be aware that "=" and "<-" are not always interchangeable in R
```

(Figure 1.1).

Basic Data Analysis for Time Series with R, First Edition. DeWayne R. Derryberry.
© 2014 John Wiley & Sons, Inc. Published 2014 by John Wiley & Sons, Inc.

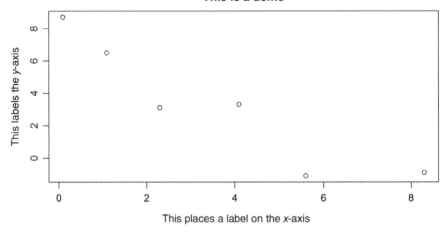

FIGURE 1.1   An example of a plot in R.

The R code produces the data plotted in Figure 1.1.

[The commands for plots used in the book are simple commands that produce bitmap files that can easily be pasted into any report and resized. The actual graphs produced for the book use the slightly more complex *tiff( )* function for better quality. For more information on better graphs check out help(tiff)].

Naming conventions used in this book:

n	is always the sample size.
y	variants represent response variables.
noise	simulated white noise (independent normal random variables).
error	random error added to models (may not be white noise).
x	variants are independent/explanatory variables.
i	the square root of -1, in other words, we will not use this for any other purpose in this book. We will not use it as a variable or name.
time	usually integers from 1 to n, representing equally spaced time intervals.

A number of simple statistical operations are worth mentioning, using the data simulated for Figure 1.1.

mean(x) produces the output: 3.583333 which is the mean of the *x* values.

sum(x) produces the output: 21.5 as expected (21.5/6 is the mean).

var(y) produces the output: 15.28667, the sample variance of *y*.

Incidentally, it is best to write code in notepad or some other text editor and then paste it in R. As the code gets longer, this becomes mandatory (more sophisticated environments exist for writing and debugging R code, such as RStudio, but they are not necessary).

1.2   SPECIAL R CONVENTIONS

Inf	infinity
NA	missing values
NaN	not a number, often an indeterminate form (Inf – Inf, 0/0, etc.)
TRUE	set a condition to true
FALSE	set a condition to false
==	Are two things equal?

Examples

2==2 will produce the output: TRUE
2=2 will produce an error, as this is an attempt to assign a value to a number
2==3 will produce the output: FALSE
1/0 will produce the output: Inf
Inf/Inf will produce the output: NaN
1 - NA will produce the output: NA

1.3   COMMON STRUCTURES

All programming languages have some looping feature and some conditional action feature. In R, the "if" statement is used for conditional action and the "for" statement for iterating a process several times. These are not the only structures in R for this purpose, but they are all that is required for this book.

The "if" statement allows an action when a condition is met:

```
x <- 2      y <- 3      z <- 2      c <- TRUE      d <- TRUE
if(x == y) {c <- 1-c}
if(x == z) {d <- 1-d}
```

Can you guess what the values of c and d will be at the end? The variable c still has the value TRUE; it was unchanged, but d has the value 0, because it is 1—TRUE, which is 0 (see Exercise 1). TRUE can be understood as 1 and FALSE can be understood as 0 in R.

The "for" statement allows an action to occur a pre-set number of times. With mathematics, the "for" statement is often a summation or multiplication action.

Consider the formula $\sum_{j=1}^{10} j^2$ which can be evaluated in closed form $m(m + 1)(2m + 1)/6$, where $m = 10$, is 385.

```
sum <- 0
for(j in 1:10)  {sum <- sum + j^2}
```

In the end, the sum will have the value 385. As with summation in most textbooks, as far as possible the letters $j, k, l,$ and m will be used when indexing summation (*i* and *n* are already used as mentioned previously), although any letter could be used. For example,

```
sum <- 0
for(z in 1:10)  {sum <- sum + z^2}
```

gets the right answer but would be considered poor style.

1.4   COMMON FUNCTIONS

pi   is the value of pi, presumably to the decimal capacity of the computer.
sqrt( )   the square root of a number. For example, sqrt(9) is 3.
log( )   the (natural) logarithm of a number. For example, log(3) is slightly more than 1.
d^c   is the base *d* raised to the power *c*. For example, 2^4 is 16.
exp( )   is exponentiation, the inverse of *log( )*. For example, log(exp(5)) is 5. Also, exp(4) is about 2.718282^4 or about 54.6.
sin( )   the sine function (all trigonometric functions assume radian inputs).
asin( )   the arcsine or inverse sine function. For example, sin(asin(4)) = 4.
cos( )   the cosine function.
acos( ) the arccosine or inverse cosine function.
tan( )   the tangent function. For example tan(3) is the same as sin(3)/cos(3).
atan( )   the inverse tangent function.

1.5   TIME SERIES FUNCTIONS

If what follows sounds like a foreign language, good, you are in the right class. Over the course of the book, these will all be introduced, but the students will learn to code primitive versions of these functions before they are allowed to use them.

acf( )   This function produces the autocorrelation plot and stores all information related to the autocorrelation function.
arima.sim( )   This function simulates random noise with a user-specified correlation structure.
ar.yw( )   This function produces all the information associated with Yule–Walker estimates for AR(*m*) processes. The related function *ar.mle( )* will also be discussed.
spec.pgram( )   This function produces a periodogram for a time series.
pacf( )   This function produces the partial autocorrelation plot and stores all information related to the partial autocorrelation function.
ts( )   This function defines a variable to be a time series.

In some sense, a substantial portion of this course is to learn these functions, and all the background associated with these functions. Throughout the book, the underlying formulas will be derived and some crude R code that produces the computations and/or plots will be presented. Only when this is done does the student know exactly what the function does and is the student permitted to use the function, which can be viewed as elegant, often cosmetic, improvements over the crude R code.

1.6   IMPORTING DATA

Obviously, data sets are often very long and manual entry is unrealistic. The functions *scan()* and *read.table()* allow for reading large files. These will be discussed in Chapter 3 when the first time series data are read in and plotted.

Only *scan()* and *read.table()* are required for this book. However, *read.csv()* and *read.delim()* can be useful for reading formats that use commas or tabs to separate items. Using library(foreign), it is possible to import data from several other software formats (SPSS, Minitab, etc.).

The next chapter is a review of simple regression and an introduction to more R code in that context.

EXERCISES

1. Guess the output of the following bits of R code and check your answers (by typing the code in R).

    (a) TRUE = FALSE; (b) TRUE == FALSE; (c) NA-NA; (d) 1/Inf; (e) Inf - Inf; (f) 1^Inf; (g) y <- 2, x <- 3, y == x; (h) 0/0; (i) 1 − TRUE; (j) 1 − FALSE; (k) TRUE + FALSE; (l) TRUE/FALSE; (m) 0^0.

2. Recall $\sum_{j=1}^{m} j = m(m+1)/2$. Write R code to find the sum $\sum_{j=1}^{20} j$ and verify that the sum is correct.

3. Write R code, using a "for" loop, to find the value 20! (20 factorial) and verify that it is correct.

4. Add $\log(15) + \tan(\pi/2)$ in R.

2

REVIEW OF REGRESSION AND MORE ABOUT R

2.1 GOALS OF THIS CHAPTER

The purpose of this chapter is threefold: (i) to review many basic notions from simple regression (the linear regression model, ordinary least squares [OLS], and the central limit theorem—in this context, basic inference); (ii) to introduce some more advanced features of R (matrix commands, curve fitting, plotting, and "inquiry" functions); and (iii) to introduce the idea of simulating data.

Real data is very important in statistics, but so is simulated data. Simulated data has known characteristics, allowing the student/programmer to examine the performance of algorithms, plots, and formulas in the best- and worst-case scenarios. Simulating data based on formulas and models allows the student/programmer to operationalize the formulas and models, often leading to a more complete understanding of what the formula or model is "saying." The ability to simulate data allows the student/programmer to quickly check conjectures and produce useful examples and counter examples. It is the opinion of the author that the ability to *effortlessly and routinely* simulate data is a skill all statisticians should have.

2.2 THE SIMPLE(ST) REGRESSION MODEL

2.2.1 Ordinary Least Squares

Imagine data produced by the following simple model: $y_k = \beta_0 + \beta_1 x_k + \varepsilon_k$.

The errors, ε_k, are normal, have mean zero, have equal spread, and are independent.

Basic Data Analysis for Time Series with R, First Edition. DeWayne R. Derryberry.
© 2014 John Wiley & Sons, Inc. Published 2014 by John Wiley & Sons, Inc.

Note: A key difference between a traditional statistical problems and a time series problem is that often, in time series, the errors are not independent.

The model can be estimated using OLS. Find estimates of β_0 and β_1, denoted $\hat{\beta}_0$ and $\hat{\beta}_1$, by solving the following problem: min $\sum (y_k - \hat{\beta}_0 - \hat{\beta}_1 x_k)^2$.

2.2.2 Properties of OLS Estimates

If the errors satisfy the assumptions of the model, these estimators are the maximum likelihood estimators and are based on the sufficient statistics.

Even when the normality assumption is not met, if the distribution of the errors has finite variance and the other assumptions are met, the estimators are BLUES—Best Linear Unbiased Estimators for the unknown parameters (Gauss–Markov theorem).

2.2.3 Matrix Representation of the Problem

The vectors for x and y can be used to form the system $\begin{pmatrix} y_1 \\ \ldots \\ y_n \end{pmatrix} = \begin{pmatrix} 1 & x_1 \\ \ldots & \ldots \\ 1 & x_n \end{pmatrix} \begin{pmatrix} \hat{\beta}_0 \\ \hat{\beta}_1 \end{pmatrix}$

or $y = X\hat{\beta}$. The solution, using linear algebra, is min $||y - X\hat{\beta}||$, has solution $X'y = X'X\hat{\beta}$, and $(X'X)^{-1} X'y = \hat{\beta}$.

Fortunately, all of this is done by R using the function *lm()* [for linear model]. However, later more customized solutions will be required and an understanding of the underlying linear algebra is needed.

2.3 SIMULATING THE DATA FROM A MODEL AND ESTIMATING THE MODEL PARAMETERS IN R

2.3.1 Simulating Data

```
n <- 50           # the sample size, n, will be 50
x <- c(1:n)       # the x vector will just be 1,2,3,...50
# simulate n = 50 random normal errors with mean 0 and standard deviation 3
error <- rnorm(n, 0, 3)
# the model is y = 3 + 0.5*x + random error (white noise in this case)
y <- 3 + 0.5*x + error
plot(x,y,pch="*") # make a scatterplot of what was simulated
title("The line y = 3 + 0.5x, n = 50")
```

The R code produces the data plotted in Figure 2.1.

2.3.2 Estimating the Model Parameters in R

2.3.2.1 The R Function lm() The OLS procedure in R is called *lm()*: fit <- lm(y ~ x).

Any name could be used to store the information from *lm()*. In this book, the convention will be to use some variant of "fit." A lot of information has been stored

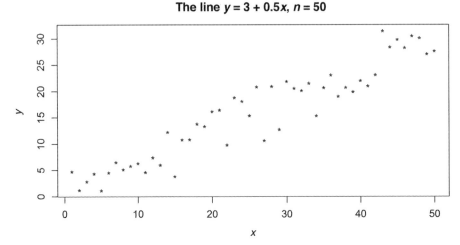

FIGURE 2.1   Simulated data for a simple regression model.

in "fit" and shortly there will be a discussion of how to access it all. The general form of the command, in pseudo code, is

"the name you choose" <- lm(response variable ~ explanatory variables).

2.3.2.2   "Inquiry" Functions in R   Important functions that are useful in R for a wide range of circumstances include *anova()*, *summary()*, *help()*, and *names()*. The commands anova(fit), summary(fit), and names(fit) all provide clues as to the information stored in "fit."

anova(fit) produces an anova (analysis of variance) table, if one exists, for any object:

anova (fit)
Response: y

	Df	Sum Sq	Mean Sq	F value	Pr(>F)
x	1	2863.67	2863.67	321.55	< 2.2e-16
Residuals	48	427.48	8.91		

Since there are many types of objects in R, it is not possible to know all the information contained in any object. The function *names()* gives us information about all that is available.

names(fit) produces a list of all the items associated with the object "fit." The items can be used/accessed using the extension fit$xxx. names(fit)

"coefficients" "residuals" "effects" "rank" "fitted.values" "assign" "qr" "df.residual" "xlevels" "call" "terms" "model".

Data with the least squares line

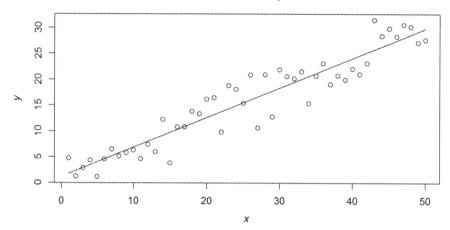

FIGURE 2.2   A least squares line fitted to the data from Figure 2.1.

For example, consider this plot, producing Figure 2.2:

```
plot(x,y)                        # plot the original data
lines(x,fit$fitted.values)       # add a line with the fitted values
title("Data with the least squares line")
```

It is important to show two other bits of code that would produce the same result:

```
plot(x,y)            # plot the original data
lines(x,fit$f)       # add a line with the fitted values
```

This demonstrates the general principle that it is only required to type a sufficient number of letters so that any item listed in *names()* can be distinguished from the other items listed.

```
predict <- fit$coeff[1] + fit$coeff[2]*x
plot(x,y)
lines(x,predict)
```

Since fit$coeff[1] is the estimated intercept and fit$coeff[2] is the estimated slope, predict = fit$fitted.values.

What are all the other things listed under names(fit)? By typing help(lm), it is possible to get a relatively complete documentation on the function *lm()*, including what all those oddball items like "fit$qr" actually are.

Finally, objects often have relatively obvious summary information. Consider the command summary(fit):

Call:		lm(formula = y ~ x)			
Residuals:					
Min	1Q	Median	3Q	Max	
−8.1788	−1.7623	0.2478	1.8572	6.4874	
Coefficients:					
	Estimate	Std. Error	t value	Pr(>|t|)	
(Intercept)	1.88170	0.85690	2.196	0.033	
x	0.52443	0.02925	17.932	<2e-16	

Residual standard error: 2.984 on 48 degrees of freedom
Multiple R-squared: 0.8701,   Adjusted R-squared: 0.8674
F-statistic: 321.5 on 1 and 48 DF, p-value: < 2.2e-16.

This is typical regression output.

Four "inquiry" functions—*names(), anova(), summary(), and help()*—have been discussed. The students should feel empowered to use these functions whenever they are curious (and be curious). When it comes to learning a programming language, some students spend too much time typing and too little time thinking, while others spend too much time thinking and too little time typing. A liberal use of these functions can be very instructive.

Functions build on top of functions in R. Consider names(summary(fit)):

"call" "terms" "residuals" "coefficients" "aliased" "sigma" "df" "r.squared".
"adj.r.squared" "fstatistic"   "cov.unscaled".

What is all this stuff? Beginning with help(summary), it becomes apparent that there is the phrase "summary.lm," so it is natural to type help(summary.lm), where the documentation for all of the items in names(summary(fit)) can be found. Explore.

2.4   BASIC INFERENCE FOR THE MODEL

Based on this model and summary(fit), the estimated intercept, $\hat{\beta}_0$, is 1.88 with an interval estimate for β_0 of roughly 0.17 (1.88 − 2·0.857) to 3.59 (1.88 + 2·0.857). Of course, the data was simulated so it is known that $\beta_0 = 3.0$, and 3.0 is indeed in the confidence interval.

Similarly, $\hat{\beta}_1$ is 0.524 with an interval estimate for β_1 of roughly 0.47 (0.524 − 2·0.0293) to 0.58 (0.524 + 2·0.0293). As before, the true value, known from the simulation code, of $\beta_1 = 0.5$ is in the interval.

The empirical rule was used in the computation of the intervals above; why can this be done? The sampling distribution of both $\hat{\beta}_0$ and $\hat{\beta}_1$ is approximately normal for large sample sizes when the errors are independent, have equal spread, and there are no outliers in either the x or y vector. The errors need not be normal, as the central limit theorem governs this process.

In the typical regression textbook, it is shown that the estimates $\hat{\beta}_0$ and $\hat{\beta}_1$ are both different linear combinations of the y values (with weights that depend on x). A linear combination of normal random variables is also normal and if the mean, variance, and covariance structure are known, the mean and variance of the resulting random variable can be determined. The typical version of the central limit theorem discussed in introductory textbooks involves identically distributed random variables combined with equal weight for each y value. If the y values are not normal but have finite mean and variance, then slight generalizations of the central limit theorem can be used to talk about weighted averages of y values as long as no particular value gets too much weight (often called leverage or influential points) (e.g., Neter et al., 1989, Chapter 3; Ramsey and Schafer, 2002, Chapter 8, Section 2).

How the standard errors were determined: The sample size is 50, the standard deviation of the residuals is 2.98, the mean of the x values is 25.5, and the variance of the x values is 212.5.

The standard error of $\hat{\beta}_1$ is $\hat{\sigma}\sqrt{\dfrac{1}{(n-1)s_x^2}} = 2.98\sqrt{\dfrac{1}{49 \cdot 212.5}} = 0.0292$.

The standard error of $\hat{\beta}_0$ is $\hat{\sigma}\sqrt{\dfrac{1}{n} + \dfrac{\bar{x}^2}{(n-1)s_x^2}} = 2.98\sqrt{\dfrac{1}{50} + \dfrac{25.5^2}{49 \cdot 212.5}} = 0.856$.

2.5 RESIDUALS ANALYSIS—WHAT CAN GO WRONG...

Once a model has been fitted to the data, the deviations from the model are the residuals. If the model is appropriate, then the residuals mimic the true errors. Examination of the residuals often provides clues about departures from the modeling assumptions.

Lack of fit—if there is curvature in the residuals, plotted versus the fitted values, this suggests there may be whole regions where the model overestimates the data and other whole regions where the model underestimates the data. This would suggest that the current model is too simple relative to some better model.

Outliers—inferences are based on the central limit theorem. If there are outliers in the data, it is possible that the y values do not have a finite variance. In this case, the central limit theorem does not apply and any inferences are potentially nonsense. If the residuals are plotted versus fitted values, outliers appear as values far from the rest of the data.

Normality—although data need not be normal, certain deviations from normality can cause concern. A probability plot [*qqnorm( )*, see Figure 2.3] of the residuals would ideally have all of the residuals falling, more or less, on a straight line. The curvature in the plot indicates skewness, and values at the ends of the data that are far from the rest of the data indicate outliers.

Influential points (outliers in x)—if there are influential points in the data, some y values have huge weight in the overall estimates of $\hat{\beta}_0$ and $\hat{\beta}_1$. This can cause either the central limit theorem to not apply or its effects to take a very large sample size to occur.

Nonwhite noise error—if the residuals appear serially correlated, great. That is what this book is about. However, the formulas from this section would not apply. As it turns out, it is hard for the untrained eye to spot serial correlation in a plot of the residuals versus order (serial correlation only makes sense if the data was collected

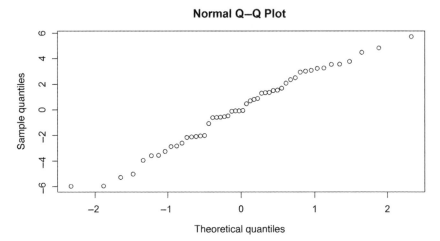

FIGURE 2.3   This plot is used for assessing violations of assumptions.

in temporal or spatial order). By the end of this book, students will know all about when serial correlation is present or absent in a set of data.

Unequal variance—sometimes the variance is larger for larger fitted values and smaller for smaller fitted values. In this case, the formulas in this section, again, do not apply. This is usually indicated by a funnel shape in the plot of residuals versus fitted values. In this context, weighted least squares are usually required.

The most useful single plot for analyzing residuals, produced in Figure 2.4, is

```
plot(fit$fitted,fit$residual)        # plot the residuals versus the fitted values
abline(0,0)                          # add a line with slope 0 and mean zero
title("Residuals plot")
```

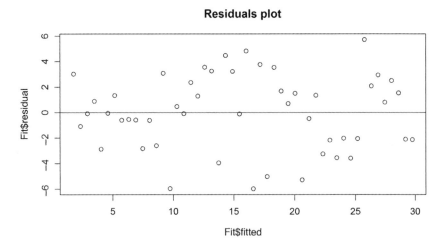

FIGURE 2.4   A plot for checking the normality of the residuals.

This is about as good a plot, with regard to assumptions, as can be found. There is no curvature, no unusual values, and no funnel shape (associated with increasing variance). Does this mean there are no problems? If there are problems with the residuals, one can be relatively certain that there are some real issues with the model, but if there are no problems with the residuals, one cannot be certain that there are no problems with the model.

A probability plot of the residuals is useful for assessing how normal the data looks, but especially for detecting outliers. Such a plot is produced in Figure 2.3 with the code

```
qqnorm(fit$resid)
```

Since the residuals follow a straight line (more or less), the residuals look close to normal. Of course, since the data were simulated, it is already known that there are no problems with the assumptions for this data.

2.6   MATRIX MANIPULATION IN R

2.6.1   Introduction

This section has two purposes, the first is to solve the regression problem as if *lm()* did not exist, and the second is to consider some other matrix- and/or vector-oriented commands.

2.6.2   OLS the Hard Way

Creating the matrix $\begin{pmatrix} 1 & x_1 \\ \dots & \dots \\ 1 & x_n \end{pmatrix}$ is as follows:

```
col_1 <- rep(1,n)         #create column 1 by repeating 1's n times
col_2 <- x                #create column 2 as just the x values
X_matrix <- cbind(col_1, col_2)# cbind() combines columns to form a matrix
 X_matrix
   col_1 col_2
[1,]   1   1
[2,]   1   2
[3,]   1   3....etc
new_fit <- solve(t(X_matrix)%*%X_matrix,t(X_matrix)%*%y)
 new_fit
col_1 2.2813698
col_2 0.5079099
```

Wow, nice solution, just like *lm()*? Recall the least squares problem in matrix form: $\min ||y - X\hat{\beta}||$, implies $(X'X)^{-1} X'y = \hat{\beta}$ from Section 2.2.3.

What has just happened? The function *t()* produces the transpose of a matrix and the operator "%*%" performs matrix multiplication. The function *solve()* solves the system $X'y = X'X\hat{\beta}$ for $\hat{\beta}$ when $X'X$ and $X'y$ are given.

2.6.3   Some Other Matrix Commands

The same X matrix could also have been created in the following manner:

```
# create a long vector with 50 1's and then the x values
x <- c(1:50)
x_temp <- c(rep(1,n),x)
# take that long collection of numbers and reshape it into a matrix
X_alt <- matrix(x_temp,50,2)          # with 50 rows and 2 columns
```

Is the matrix of the right shape? dim(X_alt): [1] 50 2
Sometimes it is useful to know the length of a vector: length(x) 50
Sometimes it is useful to access only some entries of a matrix or vector.

x[1:10] [1] 1 2 3 4 5 6 7 8 9 10 (the first 10 elements of the vector x)
mean(x[3:8]) 5.5 (the mean of the 3rd through 8th elements)
mean(X_alt[,1])   # the mean of column 1 1
mean(X_alt[3,])   #the mean of row 3 2

EXERCISES

1. An exercise for keeping in mind the true model versus the estimated model and the true error versus the estimated error (i.e., the residuals):
   Simulate linear regression data in R with white noise.
   Simulate the data for $y = 5.0 - 4.0x +$ error, experiment with the variance until the plot clearly looks linear and clearly has random noise ($0.75 < R^2 < 0.85$).
   Fit a model.
   Plot the data with both the true line and fitted line superimposed on the data.
   Find the residuals.
   Plot the residuals versus the true noise.
   Comment of the two plots created.
   Assess the assumptions with appropriate residual plots.

2. Simulate a number of regressions (2000 or more iterations) and get the sampling distribution for $\hat{\beta}$. Compare the empirical sampling distribution (mean, variance, shape) to the theory.

3. Use the *help()* function to describe the arguments that are included in the *summary()*, *matrix()*, *cbind()*, and *solve()*.

4. Create a 3 by 3 identity matrix in R.

5. Which assumption of the simple regression model does serial correlated data violate?

6. Use R to find the transpose of the matrix $\begin{bmatrix} 7 & 4 \\ 1 & 2 \end{bmatrix}$.

7. Use R to invert the matrix $\begin{bmatrix} 7 & 4 \\ 1 & 2 \end{bmatrix}$.

3

THE MODELING APPROACH TAKEN IN THIS BOOK AND SOME EXAMPLES OF TYPICAL SERIALLY CORRELATED DATA

3.1 SIGNAL AND NOISE

In general, most data can be understood as observations of the form: *signal* + *noise* = *observation* = $\widehat{signal}$ + $\widehat{noise}$. This model envisions the observations as produced by a deterministic "signal" contaminated with random "noise." In data analysis, a model is fitted to the data, producing an "estimated signal" ($\widehat{signal}$), and the resulting residuals become the "estimated noise" (*observation* − $\widehat{signal}$ = $\widehat{noise}$). The residuals, aka the estimated noise, are the basis for modeling the uncertainty in the model.

In most courses in data analysis, the focus is on white noise (independent, normal errors, with zero mean and constant variance). In the context of serial correlation, the noise itself has a complex structure based on the temporal or spatial order of the observations. The time series analysis undertaken in this book will differ from regression in that the noise has a complex structure that must be identified. Once the structure is (approximately) identified, it will be possible to perform the usual regression tasks (model selection, prediction, confidence intervals, etc.) in a routine manner, after making some relatively routine adjustments to the data. However, a substantial part of the book will involve actually developing these more complex structures for the noise and then learning how to identify their presence in data and make appropriate adjustments.

Basic Data Analysis for Time Series with R, First Edition. DeWayne R. Derryberry.
© 2014 John Wiley & Sons, Inc. Published 2014 by John Wiley & Sons, Inc.

3.2   TIME SERIES DATA

Following a long tradition, a set of observations collected over equally spaced points in time are called a time series. It should be noted that observations collected over one dimension in space, at equally spaced locations, are conceptually no different than time series, although calling them time series is a slight abuse of language. It should also be noted that time series only require the special methods discussed in this book when the observations are serially correlated.

So what is serial correlation? Serial correlation occurs when nearby observations are expected to be more similar than observations far apart (technically, serial correlation also occurs when nearby observations are more dissimilar than expected. Although this is mathematically possible, it is rare in real data). For example, if a fieldworker were to walk through the forest and take nitrogen content measurements from the soil every 20 feet, he or she would expect to see a positive association between observations close together relative to observations far apart. A keen observer of the weather will notice that temperature readings (and many other meteorological variables) are usually more similar from one day to the next, than for measurements taken several days apart. If we consider any major stock index, we notice that the index values are more similar from one day to the next than they are over longer periods of time.

At first, it appears there is a simple explanation for this in terms of signal. In each of these cases, there are underlying trends in the data that could account for the relative similarity of nearby observations. Since the stock market generally goes up, we expect the values today to be similar to the value tomorrow and both much higher than values decades ago and much lower than values in decades to come. We expect the temperatures two days in December to be more similar to a day in December than to a day in August.

What we will discover, however, is that even when these trends are removed (modeling signal), patterns still persist in the noise. For example, a very keen observer of weather will notice that temperatures tend to stay above average for several days at a time and below average for several days at a time even after adjusting for time of year. Serial correlation can actually be quite subtle. In this case, it is that the deviations from the trend do not vary as independent deviations, they tend to continue above the trend line and then continue below the trend line once they fall below it. This phenomenon is quite common and is modeled by adding a more complex structure to the noise.

This does suggest an interesting philosophical problem. Certainly it might seem that the serial correlation is due to an overly simple model for the weather. A truly keen observer of the weather (and the weather channel) would note that the long periods of above average and below average temperatures are often forecasted in advance. Whether the observed temperatures are serially correlated when compared to the weather forecast is an interesting question (there is an interesting discussion of weather forecasting in Nate Silvers book *The Signal and the Noise*, which is a recommended reading for many reasons) that will be here left unanswered.

Real data is very complex and, given a finite number of observations, the simple models proposed in this book are unlikely to truly capture "the true model," if there is such a thing, generating the data. There are at least three reasons for this: (i) the functional form of the true model is unknown; (ii) in sound statistical modeling, it is only practical to estimate a number of parameters equivalent to at most 10% of the sample size, and the model may be quite complex with too many unknown parameters relative to the sample size; and (iii) there are usually explanatory variables we have not even been measured that influence the observed outcomes.

Given all of this, a distinction between signal and noise might be viewed as arbitrary and it might be the case that, at least in some cases, it is hard to separate signal from noise. Nevertheless, this, perhaps artificial, distinction between signal and noise will be used throughout the book, and it will yield much success and occasional failure (judge for yourself). The underlying philosophical problem—if the true signal is known, in all its complexity, is the noise always white noise?—will not be answered here. However, the point is moot, the models for signal used in this book will never match the true signal (for the real data), so the noise will have a complex structure.

3.3 SIMPLE REGRESSION IN THE FRAMEWORK

The data analyzed will, in general, differ from the usual data analysis in that the noise will have complex patterns that must be diagnosed and modeled in order to produce valid statistical results. In particular, the trustworthiness of standard errors, and therefore prediction intervals, confidence intervals, and hypothesis tests will depend on properly modeling the noise.

Consider simple regression:

The model $y_i = \beta_0 + \beta_1 x_i + \varepsilon_i = signal_i + \varepsilon_i$ estimated with $y_i = \hat{\beta}_0 + \hat{\beta}_1 x_i + \hat{\varepsilon}_i$, where $\widehat{signal} = \hat{\beta}_0 + \hat{\beta}_1 x_i$ and $\widehat{noise} = \hat{\varepsilon}_i$. In modeling the noise, ε_i, it is assumed $y_i - \beta_0 - \beta_1 x_i = \varepsilon_i$ is stationary. A stationary time series is one that has had trend elements (the signal) removed and that has a time invariant pattern in the random noise. In other words, although there is a pattern of serial correlation in the noise, that pattern seems to mimic a fixed mathematical model so that the same model fits any arbitrary, contiguous subset of the noise.

A lot of the early portion of the book develops the candidate models. A complete data analysis will involve the following steps:

(i) Finding a good model to fit the signal based on the data.

(ii) Finding a good model to fit the noise, based on the residuals from the model.

(iii) Adjusting variances, test statistics, confidence intervals, and predictions, based on the model for the noise.

3.4 REAL DATA AND SIMULATED DATA

Both real and simulated data are very important for data analysis. Simulated data is useful because it is known what process generated the data. Hence it is known what the estimated signal and noise should look like (simulated data actually has a well-defined signal and well-defined noise). In this setting, it is possible to know, in a concrete manner, how well the modeling process has worked. For example, it is possible to compare parameter estimates with known values.

Real data has the important property of being real! Real data occurs in a context, has well-defined units, and analysis of the data results in answers to questions of interest. Proper analysis of real data results in brief, simple summaries of the data and results, so that colleagues with little statistical training can know the bottom line. If there were no real data, there would be no statisticians. (It is also the author's personal opinion that real data is always messier than simulated data, although this is a vague and difficult notion to explain or defend).

3.5 THE DIVERSITY OF TIME SERIES DATA

A good way to begin is by looking at a representative sample of the diverse data sets that come with the book's data diskette, and that will be analyzed later. In the section that follows, instructions will be given for loading data into R for creating these and similar graphs. Almost all of these data sets are also in DataMarket in the Time Series Data Library; the exact location will be given.

The monthly average temperature in New York City (Examples folder, "NY temps.txt"), presented in Figure 3.1, represents data in which the first goal will be to fit a periodic signal to the data, the fundamental period being annual. Using this model, it is possible, for example, to predict average temperature and estimate ranges of average temperature for any month. A lot of time series data (financial, economic, agricultural, meteorological), on earth, has an annual period.

Figure 3.2 involves the famous sunspot number data (Examples folder, "Wolf sunspots.txt"). An important issue here is determining the period and determining if

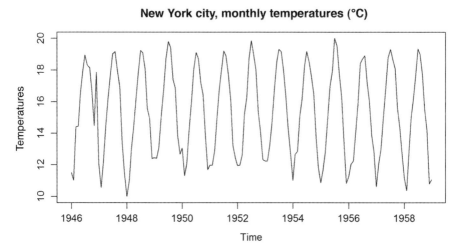

FIGURE 3.1 Monthly average temperatures of New York City (°C).

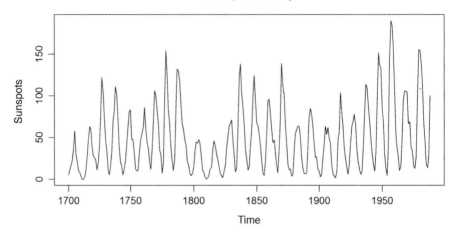

FIGURE 3.2   Wolf's Sunspot numbers, 1700–1988. Data Market–Time Series Data Library–Physics. From Tong (1991).

the period is stable or varies. Although the data is periodic, the periodic model itself may slowly change over time.

The Semmelweis data (Examples folder, "Semmelweis.txt". This data is not in the Time Series Data Library), Figure 3.3, involves an intervention. Semmelweis was a surgeon who delivered babies at a hospital around 1847. He performed a natural experiment to assess the value of sterilization of surgical instruments in surgeries. In the first 17 periods, surgeries were performed without sterilizing the instruments,

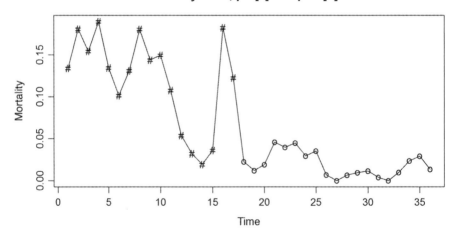

FIGURE 3.3   Semmelweis data. From Broemeling (2003).

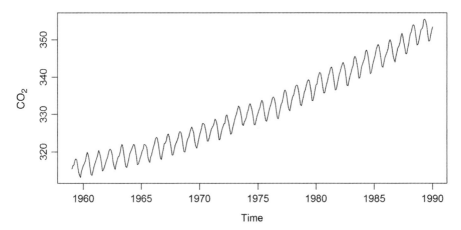

FIGURE 3.4 CO_2 (ppm) Mauna Loa, 1965–1980. DataMarket–Time Series Data Library–Meteorology. From Hipel and Mcleod (1994).

and in the last 19 periods the instruments were sterilized. The mortality rates were recorded for each period. Did the intervention reduce mortality? And if so, by how much?

Although it might appear a simple two-sample t-test with $n_1 = 17$ and $n_2 = 19$ could be used to answer the question of interest, the observations are not independent of each other. This is an egregious violation of the independence assumption underlying almost all statistical tests, including most t-tests. In this case, the two-sample t-test can be carried out, but special techniques are required in order to produce a valid test.

The CO_2 levels measured monthly from Mauna Loa (Examples folder, "MLCO2.txt"), Figure 3.4, display both a periodic pattern and a (linear?) trend. In this case a rather complex model will be fitted to the data in order to make predictions (some missing values in the data were interpolated). Although both the periodic pattern and the trend are obvious in this case, in some data sets, the starting point will be assessing whether a trend or period even exists. Model selection is required to find the "best" model from among a number of candidates. One purpose of the analysis, once a model is selected, might be to estimate the rate at which CO_2 levels are raising per year if the current pattern persists. (It should be noted that there will be limited discussion, in a book on statistical analysis, as to why the current pattern exists and may or may not persist in the future).

On the other hand, there may not be a signal (or a signal has been removed), but the noise is almost certainly not white noise (white noise = independent normal).

Figure 3.5 presents different patterns that might appear in the residuals after fitting a signal to data. It will be possible to model what is happening with the residuals in order to "correct" for these patterns. There is nothing wrong with residuals that are not white noise, but most statistical procedures assume white noise. One way to think about much of what will be done is to imagine applying linear transformations

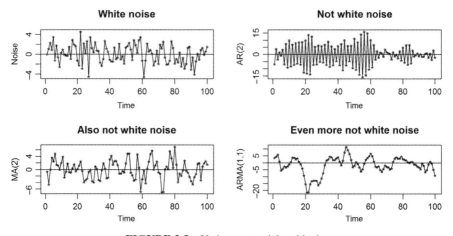

FIGURE 3.5   Various potential residuals.

to the data so that the residuals are white noise (independent), and then use standard statistical tools.

What to look for in a time series?

- The signal: trend, intervention, periodic, other?
- The structure of the random element (noise).

To reiterate: The random element in most data analysis is assumed to be *white noise*—normal errors independent of each other. In a time series, the errors are often linked so that independence cannot be assumed (the last examples). Modeling the nature of this dependence is the key to time series.

3.6   GETTING DATA INTO R

3.6.1   Overview

Data can be from many sources and appear in several formats. For this class, most of the data appears on a data diskette and can be read from there. Most of the data used in this class is originally from the Time Series Data Library compiled by Rob Hyndman. Whatever value this book may have has been greatly enhanced by the efforts of Dr. Hyndman in compiling such a rich, high-quality source of time series data in one location. The Time Series Data Library is now contained within a free source of data, DataMarket. It is important to be aware of DataMarket because it is a rich source of time series data that can be used by the instructor for additional exercises, exams, and projects.

All of the data for this class appear in one of several folders in a "*.txt" format, but may have one or more columns. If the data is made up on just one column, it can be read into R using the function *scan()*. If the data has several columns, headings, or both, then the *read.table()* function is used.

The plots that follow are presented with simple R code that produces each plot. However, DataMarket has a number of specialized tools for importing and displaying data. Additional information about getting the most out of DataMarket is included in an appendix.

3.6.2   The Diskette and the *scan()* and *ts()* Functions—New York City Temperatures

The plot for average monthly temperatures of New York City was constructed as follows.

The data is just one column of numbers, so the function *scan()* can be used. If the data diskette is on the "E" drive of your computer, the scan command would be:

```
y_NYC <- scan("E:\\Examples\\NY temps.txt")          # read the data into R,
```

R must know where to look for the file.

The plot was created by first creating a time series object with *ts()*.

```
# From the vector y_NYC, construct data with a start time of 1946 and an end time of
# 1959. Since the data is monthly, set the frequency to 12 (12 measurements per year).
temperatures <- ts(data = y_NYC,start = 1946, end = 1959, frequency = 12)
# plotting is easier because R "knows" the structure of the data
plot(temperatures)
# it is still possible to add a title
title("New York City monthly average temperature").
```

Another way to create a reasonably nice graph would have been:

```
n <- length(y_NYC)
plot(1:n,y_NYC,xlab = "month",ylab = "Temperature (C)",pch="*")
lines(1:n,y_NYC)
title("New York City average monthly temperatures")
```

Of course help(scan) and help(ts) are useful for further information.

3.6.3   The Diskette and the *read.table()* Function—The Semmelweis Data

The Semmelweis data has four columns: period, births, deaths, and mortality (which is just deaths/births). The *read.table()* command can be used here. When using *read.table()*, it must be clear whether the data has a header. The data was read into R and plotted as follows:

```
#the data has a header
y_sm <- read.table("E:\\Examples\\Semmelweis.txt", header = T)
```

Knowing how many columns there are, and how they are named requires:

names(y_sm): "intervention" "births"   "deaths"   "mortality".

The data was plotted with different symbols for observations in the pre- and post-intervention period.

```
time <- c(1:36)
# plot the data, but do not add any points
plot(time, y_sm$mort,xlab = "Period", ylab = "Mortality",pch = " ")
lines(time, y_sm$mort)
# mark the first 17 observations with one symbol (#)
points(1:17,y_sm$mort[1:17],pch="#")
# mark the last 19 observations with a different symbol (o)
points(18:36,y_sm$mort[18:36],pch = "o")
title("Childbirth mortality rates, pre[#] and post[o] intervention")
```

3.6.4   Cut and Paste Data to a Text Editor

As already mentioned, working with DataMarket, many commands are available to load data quickly and efficiently. Some crude approaches, however, can get the job done directly. In fact, when working with any data in a standard spreadsheet format (DataMarket, for example, can be used to open any file as a spreadsheet), it is possible to cut and paste the data into Notepad or some other generic text editor and save it as a "*.txt" file. It is then usually quite straightforward to use *scan()* and *read.table()* for loading the data into R.

EXERCISES

1. Consider the following data sets that could be found in the Exercises folder of the diskette or DataMarket—Time Series Data Library.
    For each data set: (a)–(e)
    (i) Scan the data into R.
    (ii) Plot the data versus time.
    (iii) Describe what signal might fit the data. If the data appears periodic, guess what the period might be.
    Meteorology:
    (a) Daily maximum temperatures in Melbourne, Australia, 1981–1990 [Exercises folder, "Melbmax.txt"].
    (b) Monthly Southern Oscillation Index measured as the difference in sea-surface air pressure between Darwin and Tahiti, January 1882–May 1993 [Exercises folder, "oscillations.txt"].

Agriculture:

(c) Monthly milk production: pounds per cow, January 1962–December 1975 [Exercises folder, "milk production.txt"].

Ecology:

(d) Annual number of lynx pelts (Hudson's Bay Company, Canada), 1857–1911 [Exercises folder, "LYNX.txt"].

Miscellaneous:

(e) Monthly average daily calls to directory assistance, January 1962–December 1976 [Examples folders, "directory assistance.txt"].

2. Simulate and store 500 observations of white noise. Using whatever name you gave these, suppose the name is "z," run the functions *acf(z)*, *pacf(z)*, *spec.pgram(z)*. This will produce three plots associated with this simulated data (do not worry too much that you do not know what you are doing).

3. Beginning to think about specialized plots.

For the same data sets from Exercise 1, run the R procedures *pacf( )*, *acf( )*, and *spec.pgram( )* and describe how the graphs differ from the similar white noise graphs in Exercise 2 (do not expect to be good at this yet!).

4

SOME COMMENTS ON ASSUMPTIONS

4.1 INTRODUCTION

Much of what is said here can be derived from any of the many introductory statistics books by David Moore et al. (for example, Baldi and Moore, 2012) and/or "The Statistical Sleuth" by Ramsey and Schafer (2002), although no one seems to have written it all in one place.

A wide variety of statistical procedures (regression, *t*-tests, ANOVA) require three assumptions:

(i) Normal observations or errors.

(ii) Independent observations (or independent errors, which is equivalent, in normal linear models to independent observations).

(iii) Equal variance—when that is appropriate (for the one-sample *t*-test, for example, there is nothing being compared, so equal variances do not apply).

These assumptions provide a minimal set of conditions required to derive a formula, whether that formula be a test statistic with a known distribution or a confidence interval. This is a derivation and is an exercise in pure mathematics. However, once that formula has been derived, the applied mathematician quickly asks: are the assumptions actually necessary for the formulas to perform properly? This is a key insight; the assumptions required for deriving any formula may or may not be crucial for using the formula.

Basic Data Analysis for Time Series with R, First Edition. DeWayne R. Derryberry.
© 2014 John Wiley & Sons, Inc. Published 2014 by John Wiley & Sons, Inc.

What is meant by "performs properly?" A 95% confidence interval is behaving properly if, in the long run, over many distinct random samples from some population, it contains the true population parameter about 95% of the time. A test statistic always produces a p-value. A test statistic is performing properly if it produces p-values that are approximately uniform$(0,1)$ when the null hypothesis is true and p-values that are right skewed when the alternative hypothesis is true.

Often, those with little applied experience believe a good strategy would be to use a statistical test to check the validity of the assumptions used to perform the test. For example, suppose someone wants to perform a two-sample t-test assuming equal variance. Although there is no clear way to check the independence assumption, they could perform a test to assess normality and a test to assess equal variance. This would be a bad idea for three reasons:

(i) All statistical tests require assumptions. If one were to perform a new test to assess the assumptions of the first test, they must then assess the assumptions of this second test before they can trust the results to make a claim about the initial assumptions. In fact, following this logic, one falls into an infinite regress. In particular, many of the tests of equal variance, tests based on F-statistics, are notoriously dependent on the normality assumptions and are almost useless for checks of equal variance in the real world (Ramsey and Schafer, 2002, p. 102; Baldi and Moore, 2012, pp. 449–450).

(ii) The tests may not actually require the assumptions being examined. If the formulas do not end up requiring the assumptions, it is a bit silly to perform detailed formal tests on assumptions that may not be needed.

(iii) Formal tests in statistics have an issue related to power and practical significance. The p-values of all formal tests are sensitive to sample size. A formal test is usually of the form:

Ho: the assumption is met versus Ha: the assumption is not met.

If the sample size is small (not an issue with the data sets in time series!), it may not be possible to reject the null hypothesis unless the violation is so obvious that any child could spot it from a graph. If the sample is quite large, the test may reject the null hypothesis when the data meets the assumptions quite well. For example, take any test of equal variance and simulate observations with a variance of 2.0 for the first sample and 2.01 for the second sample. For all practical purposes, these variances are close enough to perform a two-sample test assuming equal variance with this data. However, you can be assured that, with sufficiently large sample sizes, the null hypothesis (equal variance) will be rejected, *after all, the variances are not equal.*

Statisticians have come to realize that not all assumptions are equal and that assessing assumptions may be better done by inspection rather than formal statistical tests.

4.2   THE NORMALITY ASSUMPTION

This assumption is the least important one for the reliability of the statistical procedures under discussion. Violations of the normality assumption can be divided into two general forms: Distributions that have heavier tails than the normal and distributions that are skewed rather than symmetric. If data is skewed, the formulas we are discussing are still valid as long as the sample size is sufficiently large. Although the guidance about "how skewed" and "how large a sample" can be quite vague, since the greater the skew, the larger the required sample size. For the data commonly used in time series and for the sample sizes (which are generally quite large) used, skew is not a problem. On the other hand, heavy tails can be very problematic.

4.2.1 Right Skew

This can be confounded with another issue. If the data is right skewed, it is amazing how often the logarithmic transformation can make the data approximately normal. In rare cases, the square root transformation can do the same. This approach of transforming the data is optional but very useful. Transformations of right-skewed data are considered optional because the formulas perform properly without them, so why is it important?

When two formulas behave properly, it is still possible to ask which is best. If two test statistics produce p-values that are uniform$(0,1)$ when the null hypothesis is true and right skewed when the alternative hypothesis is true, it is still appropriate to ask which formulation produces the most right-skewed p-values, and hence is most likely to reject the null hypothesis, when the alternative hypothesis is true.

Similarly, when two formulas produce 95% confidence intervals that contain the true population parameter about 95% of the time, it is still possible to ask which formulation produces the narrowest confidence intervals.

Performing tests and constructing confidence intervals with logarithmically transformed data gets much better results, when the data is right skewed, based on the criteria set out in the last two paragraphs. The downside to this is that students need to understand logarithms and may need to transform the data back to the original scale to interpret results. When the logarithmic transformation helps, it will always be used in this book. *There is an extremely low upper bound on how far a person can continue in the field of data analysis without understanding logarithms and exponentiation.*

4.2.2 Left Skew

Fortunately, skew is a minor problem, because there is not much that can be done about left-skewed data.

4.2.3 Heavy Tails

When data is not normal, the reason the formulas are working is usually the central limit theorem. For large sample sizes, the formulas are producing parameter estimates that are approximately normal even when the data is not itself normal. The central limit theorem does make some assumptions and one is that the mean and variance of the population exist. Outliers in the data are evidence that these assumptions may not be true. Persistent outliers in the data, ones that are not errors and cannot be otherwise explained, suggest that the usual procedures based on the central limit theorem are not applicable. Outliers (and influential points, in regression) often require techniques beyond what will be covered in this course.

4.3   EQUAL VARIANCE

The equal variance assumption must be handled case by case and since this book deals only with two situations: the two-sample t-test with equal variance and with simple regression, only these two cases are discussed.

4.3.1   Two-Sample t-Test

It has been shown that the two- sample t-test is NOT sensitive to the equal variance assumptions unless the sample sizes are very different. In this course, whenever a two-sample t-test is performed, the sample sizes will be close to equal, so this will not be a problem for examples used in this book. A common check is possible however. It is always possible to perform both versions of the t-test for a common set of data, and verify that the two results are almost identical in terms of confidence intervals and p-values. When this is the case, debating the relative merits and consequences of the different tests is moot. It is difficult to find a data set for which the two procedures (the t-test with and without assuming equal variance) produce meaningfully different results.

4.3.2   Regression

When performing a regression there is usually a pattern, in the plot of residuals versus fitted values, if the variances are unequal. A weighting scheme can be developed to adjust for the unequal variance. In Chapters 13 and 16, a filtering scheme is developed for disentangling nonindependent errors, and a weighting scheme for unequal variance is easy to develop in this context. This is one of the general kinds of problems weighted least squares are designed to resolve.

4.4   INDEPENDENCE

If the observations/errors are not independent, the statistical formulations are *completely unreliable* unless corrections can be made. A large part of this course will be to develop models for nonindependence (autoregressive and moving average error structures), graphs to diagnose the types of errors from looking at the residuals (autocorrelation and partial autocorrelation plots), and methods of *filtering* the noise so that linear combinations of the original observations/errors are (approximately) independent.

TABLE 4.1   Right-Skewed Data

Unseeded clouds	Seeded clouds
1202.6	2745.6
830.1	1697.8
372.4	1656.0
345.5	978.0

4.5 POWER OF LOGARITHMIC TRANSFORMATIONS ILLUSTRATED

(Examples folder, "Cloud seeding.txt") Table 4.1 contains the total rainfall for 52 days that were candidates for cloud seeding, a method of possibly increasing rainfall. On 26 of those days, chosen at random, cloud seeding occurred, on the other 26 no cloud seeding occurred. The first few rows of these data are displayed in Table 4.1. Did cloud seeding increase rainfall? (This case is analyzed in great detail in Chapter 3 of "*The Statistical Sleuth*" by Ramsey and Schafer. Our focus will only be on the advantages of the logarithmic transformation).

If the data is read using y <- read.table(..., header = T), the code hist(y$Un) # the R command for a histogram produces Figure 4.1.

Use a logarithmic transformation with the code hist(log(y$Un)) to produce Figure 4.2.

Consider now the role of the transformation in the construction of confidence intervals.

An important idea is illustrated by constructing two confidence intervals, one on the original scale for the mean and an interval for the median using a logarithmic transformation. The interval for the median is quite involved in interpretation.

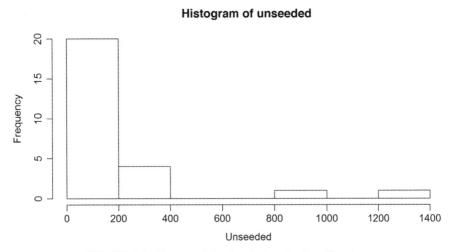

FIGURE 4.1   Extreme right skew in the cloud seeding data.

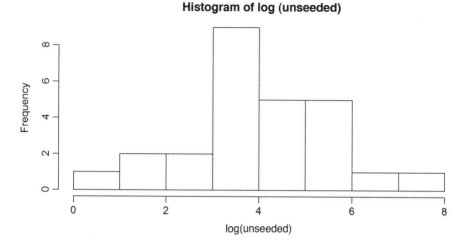

FIGURE 4.2   A logarithmic transformation removes the right skew.

t.test(y\$Un) # this function performs a t-test in R   produces a 95% confidence interval of (52.1296, 277.0473). Although *t.test()* produces a lot of output, we are only interested in the confidence interval for this example.

The median: Transformations of data alter statistics. For example, the mean of a data set can be found, but it is not easy to relate the mean of a data set to the mean of the logarithm of that data set. The median is far friendlier to transformations. If the median of a data set is found, then the logarithm of the data set is analyzed; the median of the log transformed data will be the log of the original median.

If taking the logarithmic transformation of a data set results in data that is approximately symmetric, then the mean and median are about the same, so the confidence interval for the mean [found using *t.test()*] is implicitly an interval for the median as well. This interval can be inverse transformed to produce an interval for the median on the original scale.

The function t.test(log(y\$Un)) produces a 95%confidence interval, on the logarithmic scale, of (3.327249, 4.653562). This is an interval for the mean (and median) on the log scale. So exp(3.327249) = 27.86 and exp(4.653562) = 104.96, producing a 95% confidence interval for the *median* of (27.86, 104.96).

This is an interval for the median on the original scale. This interval is lower than the interval for the mean [(27.9, 105.0) versus (52.1, 277.1)], because the median is less than the mean for right-skewed data. The interval for the median contains 44.20, which is the sample median for the data. *The interval for the median is much narrower than the interval for the mean (narrower is better).*

In general, when the data is heavily right skewed, an interval for the median on the original scale (produced by the process of taking logarithms, and inverting the resulting interval) will be lower and narrower than the confidence interval for the mean taken without using a transformation. The logarithmic transformation improves the data analysis but requires more work in interpretation.

4.6 SUMMARY

- Different assumptions differ greatly in importance.
- If observations are not independent, all hypothesis tests and confidence intervals are useless unless this problem is addressed.
- Outliers are a possible red flag, suggesting the central limit theorem may not apply in a context.
- Skew is a minor problem and transformations can often make right skew a complete nonissue.

EXERCISES

1. Compare the point estimates and interval estimates for the mean (original scale) and median (use the log scale and back transform the resulting interval) using the cloud seeding data and the seeded days.

2. Crazy Cauchy: The Cauchy distribution (aka a t distribution with one degree of freedom) is outlier prone (extremely heavy tailed). In R, Cauchy random variables can be simulated using *rcauchy()*. For example, the R code z <- rcauchy(500,0,1) will generate 500 Cauchy random variables with a median of 0 and a spread of 1 (meaning 50% of the density function lies between −1 and + 1).

    (a) Simulate 50 Cauchy random variables with median 0 and spread 1 and compute a confidence interval for the mean.

    (b) Simulate 500 Cauchy random variables with median 0 and spread 1 and compute a confidence interval for the mean.

    (c) Simulate 5000 Cauchy random variables with median 0 and spread 1 and compute a confidence interval for the mean.

    (d) Simulate 50,000 Cauchy random variables with median 0 and spread 1 and compute a confidence interval for the mean.

    (e) What might you conclude about the central limit theorem and the Cauchy distribution based on the results in (a)–(d).

5

THE AUTOCORRELATION FUNCTION AND AR(1), AR(2) MODELS

5.1 STANDARD MODELS—WHAT ARE THE ALTERNATIVES TO *WHITE NOISE*?

The two most basic alternatives to white noise are the autoregressive and moving average models for error. These are mathematical models that: (i) produce stationary noise, (ii) are based on what many view as realistic dependence structures between nearby observations, and (iii) have simple, closed form, representations. In each case there are still normal (Gaussian) errors, but there is some form of linkage between observations close in time. In this chapter the autoregressive models of order 1 and 2 [AR(1) and AR(2)] are discussed. Moving average models of order 1 and 2 [MA(1) and MA(2)] are discussed in Chapter 6. Together these ideas generalize to autoregressive models of all orders, moving average models of all orders, and autoregressive integrated moving average [ARIMA] models in Chapter 14.

Autoregression is a simple mechanism that is easy to understand and seems like a lot of what we mean by time series that are not white noise.

The simplest idea, AR(1), is just that observations closest in time have a strong correlation (in practice the correlation is almost always positive, but the general theory is the same in either case). Consider noise around a fixed mean; if the noise has an AR(1) structure with positive correlation than when an observation is above average, the next observation is likely to be above average as well, and vice-versa. For example, the weather generally displays a general pattern of this sort. If today is warmer than average, tomorrow will usually be as well, and vice-versa.

Basic Data Analysis for Time Series with R, First Edition. DeWayne R. Derryberry.
© 2014 John Wiley & Sons, Inc. Published 2014 by John Wiley & Sons, Inc.

5.2   AUTOCOVARIANCE AND AUTOCORRELATION

5.2.1   Stationarity

Any mathematical presentation of the AR(m) structures requires some simple computations based on the notion of covariance and a discussion of stationarity. A *stationary
time series (or a series in stable equilibrium)* is one in which the signal has been
removed and in which the noise displays a time invariant pattern. The series should
have a mean of zero (having removed the signal), a constant variance, and a constant
covariance structure over time. For example, any two points k units apart in the time
series should have the same covariance and correlation (some mathematicians may
quibble with this informal definition, but the rest of us can move on).

Just to make a clear distinction. Most time series books consider a series stationary
even with a constant mean. On the other hand, in linear models that mean is part of
the signal to be modeled. In this book, a stationary series always has a mean of
zero. This, otherwise arbitrary distinction, is natural when taking a regression/linear
models approach to modeling the data.

The observations in a stable time series, with the signal removed, can be denoted
$\varepsilon_1, \varepsilon_2, \ldots \varepsilon_n$ and also be thought of as errors or noise.

(i) $E(\varepsilon_j) = 0$, for all j—the signal has been removed.

(ii) $\text{Var}(\varepsilon_j) = \sigma^2$, for all j—the variance is constant.

(iii) $\text{Cov}(\varepsilon_j, \varepsilon_{j+k}) = C_k$, for all j and k—the covariance is always the same between
observations and k units apart. The integer k is the "lag." This notion of the
covariance between observations k units apart is the fundamental idea in all
later developments. This is obviously the precise point where this book begins
to deviate from a standard text on regression.

5.2.2   A Note About Conditions

Notice that the definition of stationarity used here is slightly stronger than that used in
most time series books. Furthermore, Gaussian errors will be assumed in this book.
Although care must be taken in any case to verify the hypotheses of any theorem,
any results about stationarity in other time series books would generally apply in this
context.

5.2.3   Properties of Autocovariance

The notion of stationarity is now clear. For a stationary time series these three
conditions hold, but the last condition implies a kind of stochastic time invariance:

$$\text{Cov}(\varepsilon_j, \varepsilon_{j+k}) = \text{Cov}(\varepsilon_{m+j}, \varepsilon_{m+j+k}) = C_k, \quad \text{for every integer } m.$$

An immediate implication is that $C_k = C_{-k}$ (why?), a fact that is often useful.

Furthermore, $\text{Cov}(\varepsilon_j, \varepsilon_j) = \text{Cov}(\varepsilon_{j+m}, \varepsilon_{j+m}) = C_0$, for any m. As with linear
models, a correlation, the autocorrelation, can be defined to be $R_k = C_k/C_0$.

Autocorrelation is a special case of correlation, as defined in probability, so $-1 \leq R_k \leq 1$ follows from a form of the Cauchy–Schwartz inequality.

For Gaussian processes (normal errors) the covariance structure is a complete description (Box et al., 2008, p. 28) of a stationary time series.

5.2.4 White Noise

More complex errors will be built up from white noise. White noise, w_j, is normal error with: (i) $E(w_j) = 0$, (ii) $E(w_k w_k) = \sigma_w^2$, and (iii) $\text{Cov}(w_i, w_{i+k}) = E(w_i \cdot w_{i+k}) = E(w_i)E(w_{i+k}) = 0$. From probability it is known that independence always implies zero correlation, and for multivariate normal random variables, zero correlation implies independence.

5.2.5 Estimation of the Autocovariance and Autocorrelation

So far the results have been about errors. These are continuous random variables defined by integrals (expectations and covariances). In order to estimate the auto-covariance and autocorrelation, the residuals from a fitted signal, denoted $\hat{\varepsilon}_i$, are combined using sums and averages.

When a signal is fitted to the data using regression it is usually the case that $\frac{1}{n} \sum \hat{\varepsilon}_i = 0$.

The autocovariance and autocorrelation are estimated with $\hat{C}_k = \frac{1}{n} \sum_{j=k+1}^{n} \hat{\varepsilon}_j \cdot \hat{\varepsilon}_{j-k}$ and $\hat{R}_k = \hat{C}_k / \hat{C}_0$. (That the divisor for $\hat{C}_k$ is n, rather than $n - k$, is a technical details ignored here. It should be noted that when n is large relative to k it makes little difference).

A usual assumption is that all of the autocorrelations beyond some point become nearly zero. If this is not the case, it will be impossible to model the time series with any finite sample. Furthermore, in the real world, dependence does seem to fade with time (with increasing lag k). We expect observations close in time to be very similar, but we do not expect observations far apart to be similar.

As a practical matter, a time series should usually have at least 50 observations, and one should expect to compute no more than $k \approx n/4$ autocovariances (Box et al., 2008, p. 32).

In the case of white noise, a baseline for comparisons throughout the book, it is known that $\hat{R}_k$ has asymptotic mean zero and variance $1/n$ (Box et al., 2008, p. 34).

5.3 THE *acf()* FUNCTION IN R

5.3.1 Background

It is natural to plot the autocovariances and/or the autocorrelations versus lag. Further, it will be important to develop some notion of what would be expected from such a plot if the errors are white noise (meaning no special time series techniques are required) in contrast to the situation strong serial correlation is present. Because the autocorrelations are normalized to vary between −1 and 1, this will be the default version of the plot.

In R there is a function that produces just this plot, but it will be worth it to produce such a plot from scratch before relying on R completely.

5.3.2   The Basic Code for Estimating the Autocovariance

Strategy for computing and plot *acf( )* values:

(i) Simulate or scan in the series.

(ii) Get the length of the series (get n).

(iii) Set up a storage vector for the residuals, $\hat{C}_k$ and $\hat{R}_k$.

(iv) Estimate a signal using regression.

(v) Remove a signal to create residuals and name/store the residuals.

(vi) Compute $\hat{C}_k$ and $\hat{R}_k$ [using a "for" loop with $\hat{C}_k = \frac{1}{n} \sum_{j=k+1}^{n} \hat{\varepsilon}_j \cdot \hat{\varepsilon}_{j-k}$ and $\hat{R}_k = \hat{C}_k / \hat{C}_0$].

(vii) Plot the values versus k.

Because the data here will be simulated, *scan( )* is not used. The simulated data is just white noise.

```
# simulated data
n <- 500                              # 500 observations
errors <- rnorm(n,0,4)                # errors
# remove the mean (signal)
residuals <- errors - mean(errors)
store_Ck <- rep(0,25)                 # a vector to store correlations
store_Rk <- rep(0,25)                 # a vector to store covariances
# computing and storing Ck and Rk
C_0 <- sum(residuals*residuals)/n
```

```
# the inner for loop is operationalizing the summation formula for a covariance
for (k in 1:25)
{for( j in (k+1):n)
              {store_Ck[k] <- store_Ck[k] + residuals[j]*residuals[j-k]}}
store_Ck <- store_Ck/n
store_Ck <- c(C_0,store_Ck)
store_Rk <- store_Ck/C_0
# plot the information
lag <- c(1:25)          #the choice of number of lags was somewhat arbitrary
plot(0:25,store_Rk,xlab = "lag", ylab = "autocorrelation")
title( main = "autocorrelation plot by hand")
lines(0:25,store_Rk)
```

```
abline(0,0)
# bands within which the autocorrelations should fall (about 95% of the time)
# for white noise)
abline(2/sqrt(n),0,lty = 2)
abline(-2/sqrt(n),0,lty = 2)
```

Figure 5.1 is of the autocorrelation plot constructed using definitions (on the left) versus the R command: acf(residuals) (on the right). Both display the usual plot for white noise. When time series have serial correlation, large spikes outside the bands will appear, usually above the bands (indicating positive autocorrelations) and usually large for early spikes (the strongest correlation being between nearby observations).

So what all can be learned about *acf( )*? The commands: help(acf), and z <- acf(residuals), names(z) are informative: names(z) "acf" "type" "n.used" "lag" "series" "snames".

It is easy to verify, printing z$acf and "store_Rk" that the computations are identical. So the code above and *acf( )* are different presentations of the same plot.

Periodic data will display large positive spikes at the period and may have negative spikes at the half period. For example, data gathered monthly, given an annual cycle, will have strong positive spikes at lags 12, 24, 36, so on and often have strong negative spikes at lags 6, 18, 30, so on.

The file "Accidental deaths.txt" from the Examples folder (DataMarket–Time Series Data Library–Demography–Accidental deaths in USA: monthly, 1973–1978) displays this annual pattern.

By extending the number of computed lags to 72 acf(...,lag.max = 72), it is clear (Figure 5.2) that the monthly periodic pattern, with period 12, is quite persistent in the data.

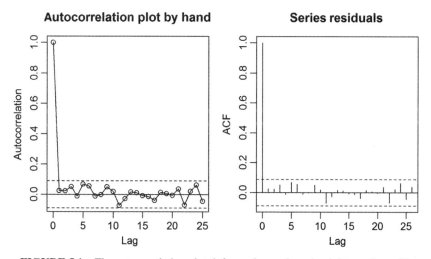

FIGURE 5.1   The autocorrelation plot: left panel—crude code, right panel—*acf( )*.

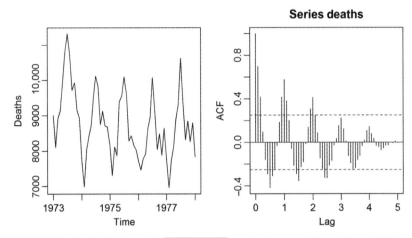

FIGURE 5.2   Periodic data, with lag.max = 72 used to increase the total lags displayed.

5.4   THE FIRST ALTERNATIVE TO WHITE NOISE: AUTOREGRESSIVE ERRORS—AR(1), AR(2)

5.4.1   Definition of the AR(1) and AR(2) Models

Let ε_j be a stationary time series as previously described. The AR(m) process is defined as $\varepsilon_j = a_1\varepsilon_{j-1} + a_2\varepsilon_{j-2} + \cdots + a_m\varepsilon_{j-m} + w_j$, where w_j is white noise with variance σ_w^2. The AR(1) and AR(2) processes are then AR(1): $\varepsilon_j = a_1\varepsilon_{j-1} + w_j$ and AR(2): $\varepsilon_j = a_1\varepsilon_{j-1} + a_2\varepsilon_{j-2} + w_j$.

For the moment the focus is on these two, because they have a number of closed form results. For a time series to be stationary, some conditions on the coefficients a_i must be met. It is easy to show that the AR(1) process is stationary when $|a_1| < 1$. On the other hand, it is quite difficult to show, but true, that the AR(2) process is stationary when (i) $a_1 + a_2 < 1$, (ii) $a_2 - a_1 < 1$, and (iii) $|a_2| < 1$.

5.4.2   Some Preliminary Facts

In the next two chapters and throughout the book several basic rules of expectation will be routinely employed. These can be found in any book on probability. However, the rules listed here are either naming conventions or rules that should be intuitively obvious from the time series context, regardless of any detailed knowledge of probability.

$E(w_k w_k) = \sigma_w^2$ is the white noise variance.

$E(\varepsilon_j \varepsilon_j) = \sigma_{AR}^2$ is the error variance when the model is AR(m).

$E(w_j w_m) = 0$ when $j \neq m$ because white noise random variables are independent.

$E(w_k \varepsilon_m) = 0$ when $m < k$ because the white noise has occurred after the error in time.

$E(w_k \varepsilon_m) = ?$ when $m \geq k$, this depends on how the AR(m) model is defined.

Recall we have $E(w_k) = 0$ and $E(\varepsilon_k) = 0$ for all k.

5.4.3 The AR(1) Model Autocorrelation and Autocovariance

Data with an AR(m) structure is simulated throughout the book, and a number of properties are well known. The formulas derived determine, for example, the autocovariances as a function of a_1 and a_2.

For the AR(1) model, $\varepsilon_j = a_1\varepsilon_{j-1} + w_j$, $\varepsilon_j = a_1^2\varepsilon_{j-2} + a_1 w_{j-1} + w_j$, and in general $\varepsilon_j = a_1^k\varepsilon_{j-k} + \sum_{t=1}^{k} a_1^{t-1}w_{j-t+1}$ (the proof is left as an exercise).

Furthermore $C_1 = E(\varepsilon_j \cdot \varepsilon_{j-1}) = E([a_1\varepsilon_{j-1} + w_k]\varepsilon_{j-1}) = a_1\sigma_{AR}^2 = a_1 C_0$.

More generally, $C_k = E(\varepsilon_j \cdot \varepsilon_{j-k}) = E([a_1^k\varepsilon_{j-k} + \sum_{t=1}^{k} a_1^{t-1}w_{j-t+1}] \cdot \varepsilon_{j-k}) = a_1^k C_0$.

So, in general $R_k = a_1^{|k|}$ [recall, by stationarity, $C_k = C_{-k}$].

For AR(1), the relationship of variance in the series to variance in white noise is

$$\sigma_{AR}^2 = E(\varepsilon_j\varepsilon_j) = E([a_1\varepsilon_{j-1} + w_j][a_1\varepsilon_{j-1} + w_j]) = a_1^2\sigma_{AR}^2 + \sigma_w^2, \text{ so}$$

$$\sigma_{AR}^2 = \sigma_w^2/\left(1 - a_1^2\right).$$

Notice that, for AR(1), $\sigma_{AR}^2 > \sigma_w^2$ in all cases. Furthermore, if $|a_1| \geq 1$, these formulas produce nonsensical results, such as negative variances. The conditions on the values a_1 and a_2 can be viewed simply as restrictions that prevent nonsensical values for variances and correlations. In Chapter 14 it will be shown more explicitly that the stationary conditions for AR(2), and all AR(m) models, are simple extensions of this idea.

5.4.4 Using Correlation and Scatterplots to Illustrate the AR(1) Model

The model AR(1) is about the correlation between an error and the previous error. This can be modeled using traditional correlations and scatterplots where the x and y variables are actually the error and the error shifted one time unit (obviously when variable is x and which is y is irrelevant here). Below is some simulated data and scatterplots of errors and shifted errors.

First consider white noise. In this case there should be no correlation between the errors and the shifted errors.

Figure 5.3 is a scatterplot of 1000 white noise errors versus the same errors shifted by one time unit. There is no pattern in the data.

On the other hand, consider AR(1) errors with $a_1 = 0.7$. Figure 5.4 demonstrates the basic dependence of each error on the previous errors.

5.4.5 The AR(2) Model Autocorrelation and Autocovariance

For the AR(2) process the formulas are not so easily formed in closed form.

The model is: $\varepsilon_j = a_1\varepsilon_{j-1} + a_2\varepsilon_{j-2} + w_j$, and the autocovariances can be characterized with recursive relationships. $C_1 = E(\varepsilon_j\varepsilon_{j-1}) = E([a_1\varepsilon_{j-1} + a_2\varepsilon_{j-2} + w_j]\varepsilon_{j-1}) = a_1 C_0 + a_2 C_1$, so it is easy to see $R_1 = a_1/(1 - a_2)$. The derivation can be generalized, $R_k = a_1 R_{k-1} + a_2 R_{k-2}$.

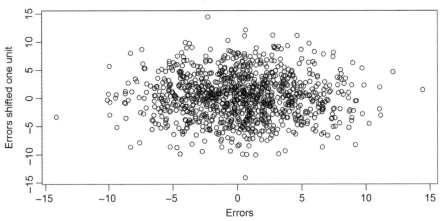

FIGURE 5.3   The correlation is 0.0065, essentially zero.

For example $R_2 = a_1^2/(1 - a_2) + a_2$ and $R_3 = (a_1^3 + a_1 a_2)/(1 - a_2) + a_1 a_2$. With great labor it can be shown that, for the AR(2) model, $\sigma_{AR}^2 = \sigma_w^2/(1 - a_1 R_1 - a_2 R_2)$ (exercise).

5.4.6   Simulating Data for AR(m) Models

The general pattern for the R code presented below is as follows:

Simulating the series.
Define length of series (make it long!).

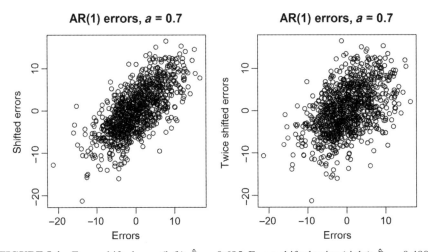

FIGURE 5.4   Errors shifted once (left), $\hat{R}_1 = 0.695$, Errors shifted twice (right), $\hat{R}_2 = 0.480$.

Define/set AR parameters.

Set up time and error storage vectors.

Simulate white noise.

Use AR definitions and white noise to produce the errors.

Verification of the results:

Plot the data.

Determine *acf()* for the data.

Example R code:

```
# Setting up the simulation
n <- 1000
time <- c(1:n)
error <- rep(0,n)
a1 <- 0.8
a2 <- -0.7              #for AR(1) just set a2 <- 0
#simulating white noise
noise <- rnorm(n, 0,2)
# use AR definitions to produce non-white noise
error[1] <- noise[1]
error[2] <- a1*error[1] + noise[2]
error[3] <- a1*error[2] + a2*error[1] + noise[3]
for(k in 4:n)
{error[k] <- a1*error[k-1] + a2*error[k-2] + noise[k]}
# plot data and assess model
plot(time, error, type = "l")
abline(0,0)
z <- acf(error)
```

Based on the formulas already derived and the choices used in the code ($a_1 = 0.8$, $a_2 = -0.7$):

$$R_0 = 1 \text{ by definition}$$
$$R_1 = a_1/(1 - a_2) = 0.8/1.7 \approx 0.4706$$
$$R_2 = a_1R_1 + a_2R_0 \approx 0.8(0.4706) - 0.7(1) \approx -0.3235$$
$$R_3 = a_1R_2 + a_2R_1 \approx 0.8(-0.3235) - 0.7(0.4706) \approx -0.5882, \text{etc.}$$

A visual inspection of Figure 5.5 shows good agreement between the simulated values and the expected values for the first four lags.

It should be noted that the first few simulated values of the series are not yet AR(2). There is a burn-in period before the series is truly AR(2), and the first few terms could be dropped from the analysis. However, the objective is to use the

Series error

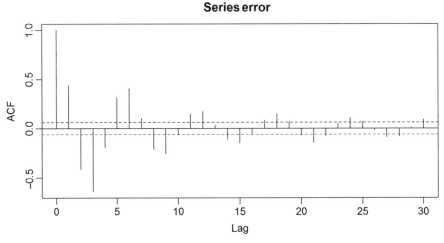

FIGURE 5.5   The *acf( )* plot for AR(2) errors when $a_1 = 0.8$, $a_2 = -0.7$.

simulation to reinforce and operationalize the definition of the AR(*m*) model. Later a function will be introduced that simulates AR(*m*) errors (and many other structures) directly.

5.4.7   Examples of Stable and Unstable AR(1) Models

Consider the following three cases of AR(1) data: $a_1 =$ (i) –0.9, (ii) +0.5, (iii) +1.01

(i) The first case (Figure 5.6), $a_1 = -0.9$, produces a strong negative correlation between nearby observations (an unrealistic scenario in most real situations, but theoretically possible). This kind of random noise "bounces" above and below zero more than white noise.

(ii) The second case (Figure 5.7), $a_1 = 0.5$, produces a more realistic positive correlation between nearby observations; the correlation is also weaker than in the previous case. This kind of random noise "bounces" above and below zero less often than white noise.

(iii) The last case (Figure 5.8), $a_1 = 1.01$, violates the requirements of stable equilibrium only slightly, yet will display long-run instability. At some point the errors will produce exponential growth.

In Figure 5.6, the data crosses the center line more than white noise. The *acf( )* function alternates between positive and negative correlations at odd and even lags for several lags, then become indistinguishable from white noise. The theoretical autocorrelations are $R_1 = -0.9$, $R_2 = 0.81$, $R_3 = -0.73$, so on, while the observed values are –0.829, 0.683, –0.565, so on.

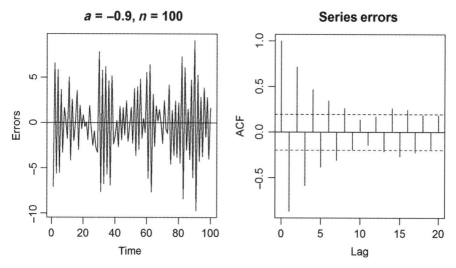

FIGURE 5.6   Case (i). The data (left) and the *acf( )* plot (right).

In Figure 5.7, the data crosses the center line less than white noise. The autocorrelation values are all positive for the first few lags, then late lags are indistinguishable from white noise. Because the sample size is the same as the last example, but the value of a_1 is smaller in magnitude, the theoretical autocorrelations collapse to white noise-like values more quickly. The theoretical values are $R_1 = 0.5$, $R_2 = 0.25$, and $R_3 = 0.125$, so on, while the sample values are 0.563, 0.269, and 0.151, so on.

For a stationary model, the errors must average zero. Figure 5.8 displays data that does not average to zero, even though no signal is involved. From a practical point of

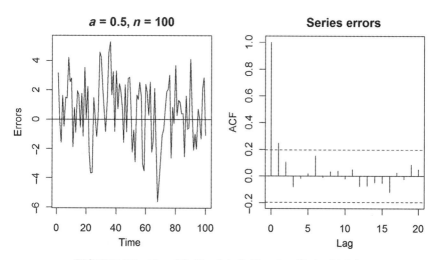

FIGURE 5.7   Case (ii). The data (left) and *acf( )* plot (right).

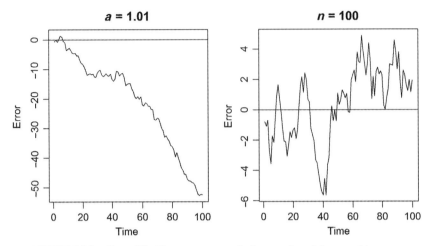

FIGURE 5.8   Case (iii). These are two typical examples of the unstable case.

view, it should be obvious that any model with a signal and this kind of noise could not be realistically modeled using statistical tools and a finite sample size. In an exercise the reader is asked to simulate several series of data from an unstable model and plot them all together. This will give a clear sense of how instability behaves in general, rather than in one instance.

5.4.8   Examples of Stable and Unstable AR(2) Models

Recall the AR(2) process is stationary when (i) $a_1 + a_2 < 1$, (ii) $a_2 - a_1 < 1$, and (iii) $|a_2| < 1$. Table 5.1 outlines 10 stationary and unstable choices for a_1 and a_2.

In Chapter 14, it will be shown that the model simulated in Figure 5.9 is equivalent to two applications of the AR(1) model with positive a_1 in both cases. The sample values are $\hat{R}_1 = 0.929$, $\hat{R}_2 = 0.822$, and $\hat{R}_3 = 0.685$.

In Chapter 14, it will be learned that the data in Figure 5.10 is from a model equivalent to two applications of the AR(1) model with a positive and a negative a_1,

TABLE 5.1   Examples of Stable and Unstable AR(2) Models

Case	a_1	a_2	R_1	R_2	R_3
(iv) Stable	1.60	−0.63	0.982	0.941	0.887
(v) Stable	0.30	0.40	0.500	0.550	0.365
(vi) Stable	−0.30	0.40	−0.500	0.550	−0.365
(vii) Stable	1.20	−0.72	0.698	0.118	−0.361
(iix) Unstable violates (i)	0.500	0.501	NA	NA	NA
(ix) Unstable violates (ii)	−0.51	0.500	NA	NA	NA
(x) Unstable violates (iii)	0.0	−1.01	NA	NA	NA

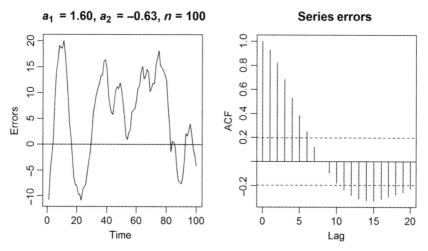

FIGURE 5.9   Case (iv). Equivalent to two iterations of AR(1) with positive a.

the positive value being larger. The sample values are $\hat{R}_1 = 0.409$, $\hat{R}_2 = 0.460$, and $\hat{R}_3 = 0.199$.

The model represented in Figure 5.11 is equivalent to two applications of the AR(1) model with a positive and a negative a_1, but now with the negative value being larger. The sample values are $\hat{R}_1 = -0.348$, $\hat{R}_2 = 0.321$, and $\hat{R}_3 = -0.038$.

The data from Figure 5.12 is produced by a model equivalent to two applications of the AR(1) model, where the a_1 values are complex conjugates (for those unfamiliar with complex numbers, this will be explained in Chapter 7). The sample values are $\hat{R}_1 = 0.684$, $\hat{R}_2 = 0.057$, and $\hat{R}_3 = -0.453$.

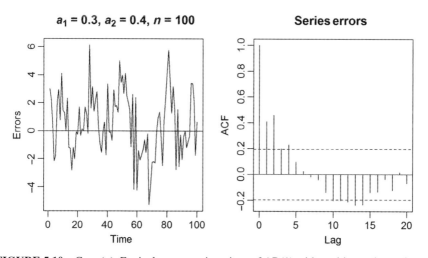

FIGURE 5.10   Case (v). Equivalent to two iterations of AR(1) with positive and negative a.

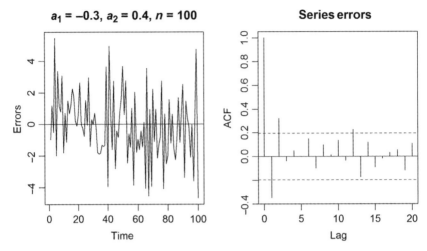

FIGURE 5.11   Case (vi). Equivalent to two iterations of AR(1) with positive and negative a.

Before displaying the unstable cases, the reader may have noticed that the sample autocorrelation values are usually less than the theoretical values. Although the sample estimates of the autocorrelations are asymptotically unbiased (any bias converges to zero with large sample size), the estimates are not unbiased. If fact, the division by n in the denominator of the sample autocorrelation formula (Section 5.2.5) tends to slightly underestimate the true value. This bias is more pronounced for that larger lags.

In the last two cases (Figure 5.13) it seems to be the increasing variance, not the move from the zero average, that is the source of the instability (recall stability

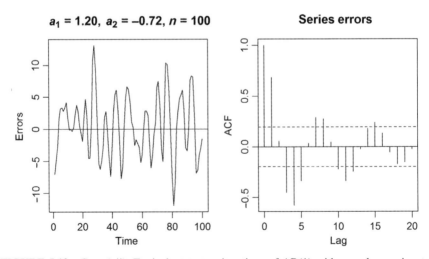

FIGURE 5.12   Case (vii). Equivalent to two iterations of AR(1) with complex conjugates for a.

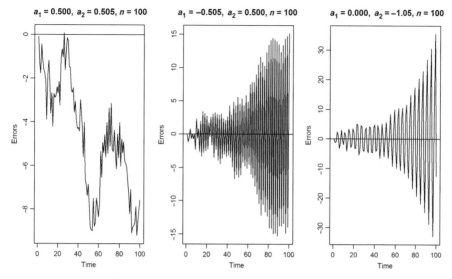

FIGURE 5.13   Unstable cases (iix), (ix), and (x).

implies time invariance, a time series that has a growing variance fails to display a pattern invariant over time). Further study of these instable cases might be a topic in a class in stochastic processes and might be quite interesting, but this book will focus on the more mundane cases encountered in real life.

EXERCISES

1. Write the code to compute the *acf()* plot yourself for the file "Furnas 31 78.txt" from the Exercises folder (DataMarket–Hydrology–Monthly riverflow in cms, Furnas–vazoes medias mensais, 1931–1978). Also use the R function *acf()* for comparison. Since the data is monthly you should compute 4–5 years of lags to check for a persistent periodicity. Is the period obvious in the plot? Explain.

2. For the lynx data, "LYNX.txt", from the Exercises folder (DataMarket–Ecology–Annual number of lynx trapped, MacKenzie River, 1821–1934):

   (i) Plot the data versus time as a time series.

  (ii) Does the data appear periodic? If so, what it the period?

  (iii) Use R to compute an *acf()* plot for the series.

  (iv) Comment on what you found.

3. Show that when $a_2 = 0$, the stability conditions for an AR(2) model are equivalent to the stability conditions for an AR(1) model.

4. Show that for the AR(1) process: $\varepsilon_n = a_1^k \varepsilon_{n-k} + \sum_{j=1}^{k} a_1^{j-1} w_{n-j+1}$.

5. For an AR(1) process, $\varepsilon_n = a_1^2 \varepsilon_{n-2} + a_1 w_{n-1} + w_n$ (the formula from Exercise 4 with $k = 2$). Use this fact to re-derive the formula $\sigma_{AR}^2 = \sigma_w^2/(1 - a_1^2)$.

6. Show that, for the AR(2) model, $\sigma_{AR}^2 = \sigma_w^2/(1 - a_1 R_1 - a_2 R_2)$. hint: expand $E(\varepsilon_j \varepsilon_j) = E([a_1 \varepsilon_{j-1} + a_2 \varepsilon_{j-2} + w_j]^2)$ and exploit the similarities between these formulas.

7. Solve for the values in the following table (with $\sigma_w^2 = 4$)

a_1	a_2	R_1	R_2	R_3	σ_{AR}^2
??	??	0.8	0.6	??	??
		0.4	-0.1		
		-0.3	0.7		
??	??	-0.5	-0.4	??	??

8. Simulate each series in the previous table (Exercise 7) with the specified white noise variance:

   (i) Find a_1 and a_2 and verify they meet the conditions for stability.

   (ii) Compute all values on the table.

   (iii) Simulate and plot the time series.

   (iv) Produce the *acf()* graph.

   (v) Verify R_1, R_2, R_3 and the variance for the series [use var() in R to get the variance of the time series].

9. Write an R program plot several (perhaps 50 or 100) simulated AR(1) series with $a = \pm(1 + \varepsilon)$, where ε is some small number. Describe the general pattern suggested.

6

THE MOVING AVERAGE MODELS MA(1) AND MA(2)

6.1 THE MOVING AVERAGE MODEL

The second commonly used model for temporal patterns in the errors is the moving average structure. The structure MA(l) is $\varepsilon_j = -b_l w_{j-l} \ldots - b_2 w_{j-2} - b_1 w_{j-1} + w_j$, where w_k are all white noise. In this chapter the focus is on the special cases MA(1) and MA(2). As with AR(m) models, the general MA(l) cases are dealt with in Chapter 14.

6.2 THE AUTOCORRELATION FOR MA(1) MODELS

The basic model for MA(1) is $\varepsilon_j = -b_1 w_{j-1} + w_j$. Using the relationship(s) $\varepsilon_j = -b_1 w_{j-1} + w_j$, $\varepsilon_{j-1} = -b_1 w_{j-2} + w_{j-1}$, and $\varepsilon_{j-2} = -b_1 w_{j-3} + w_{j-2}$, it is easy to derive (Exercise 1) the autocorrelation function, R_k $C_0 = \sigma_{MA(1)}^2 = (1 + b_1^2)\sigma_w^2$, $C_1 = -b_1\sigma_w^2$, and $C_j = 0$ for $j \geq 2$.

From this it follows that: $R_0 = 1$, $R_1 = -b_1/(1 + b_1^2)$, and $R_k = 0$ for $k \geq 2$.

There is an invertibility condition $|b_1| < 1$. The reason for this condition will be more apparent in Chapter 14. Notice that, for any real b_1, all R_k, $k = 0, 1, 2, \ldots$ are

Basic Data Analysis for Time Series with R, First Edition. DeWayne R. Derryberry.
© 2014 John Wiley & Sons, Inc. Published 2014 by John Wiley & Sons, Inc.

valid, so the invertibility condition is obviously not functioning in the same way as the stability conditions for AR(m) models.

6.3 A DUALITY BETWEEN MA(l) AND AR(m) MODELS

The following argument is not particularly rigorous but hints at some more general results found in Chapter 14. Recall, for the AR(1) model, $\varepsilon_n = a_1\varepsilon_{n-1} + w_n$, so $\varepsilon_n = a_1^2\varepsilon_{n-2} + a_1 w_{n-1} + w_n$ and in general $\varepsilon_n = a_1^k\varepsilon_{n-k} + \sum_{j=1}^{k} a_1^{j-1} w_{n-j+1}$. More generally, let $k \to \infty$. Because $|a_1| < 1$, then $\varepsilon_n \approx \sum_{j=1}^{\infty} a_1^{j-1} w_{n-j+1}$.

In other words, it appears that, in some sense, AR(1) $\equiv$ MA(∞). In the exercises you will be asked to perform a similar analysis that suggests, in some sense, MA(1) $\equiv$ AR(∞). A more general duality between AR(m) and MA(l) models, suggested by these intuitive "derivations," will be developed in Chapter 14.

6.4 THE AUTOCORRELATION FOR MA(2) MODELS

The model for MA(2) is $\varepsilon_j = -b_{j-2}w_{j-2} - b_1 w_{j-1} + w_j$. It is not hard to derive Var$(\varepsilon_j) = C_0 = \sigma_{MA(2)}^2 = (1 + b_1^2 + b_2^2)\sigma_w^2$, $C_1 = -b_1(1 - b_2)\sigma_w^2$, $C_2 = -b_2\sigma_w^2$, and $C_k = 0$ for $k \geq 3$. Therefore $R_0 = 1, R_1 = -b_1(1 - b_2)/(1 + b_1^2 + b_2^2), R_2 = -b_2/(1 + b_1^2 + b_2^2)$, and $R_k = 0, k \geq 3$ (these derivations, which are quite straightforward, are Exercise 2). The invertibility conditions mirror the stability conditions for the AR(2) model: (i) $b_1 + b_2 < 1$, (ii) $b_2 - b_1 < 1$, and (iii) $|b_2| < 1$.

6.5 SIMULATED EXAMPLES OF THE MA(1) MODEL

There are two cases, positive and negative values. Violations of the invertibility condition do not produce any obvious instability, so there is no need to show this case.

Case (i) $b_1 = -0.7$
(ii) $b_1 = 0.3$

The strategy for producing MA(l) data:

Define length of series (make it long!)

Define/set MA parameters.

Set up time and error storage vectors.

Simulate white noise

Use MA definitions and white noise to simulate the data.

Assess the series

    Plot the data.

    Determine *acf( )* for the data.

The R code is similar to the code for AR(m) models; expect the following lines must be changed:

```
b1 <- -0.8          # the parameters are labeled "b" instead of "a"
b2 <- 0.0           #for MA(1) just set b2 <-0
```

The definition of the errors is changed to reflect the definition of MA(l) models:

```
error[1] <- noise[1]
error[2] <- -b1*noise[1] + noise[2]
error[3] <- -b1*noise[2] - b2*noise[1] + noise[3]
for (k in 4:n)
{error[k] <- -b1*noise[k-1] - b2*noise[k-2] + noise[k]}.
```

The theoretical values for the first case (Figure 6.1) are: $R_1 = 0.47$, $R_2 = 0$, and $C_0 = 1.49\sigma_w^2$. The values, estimated from this simulated data, were 0.373, –0.243, and $1.52\sigma_w^2$.

Simulated data for the second case is found in Figure 6.2. The theoretical values are $R_1 = -0.275$, $R_2 = 0$, and $C_0 = 1.09\sigma_w^2$. The observed values being –0.111, 0.01, and $0.88\sigma_w^2$.

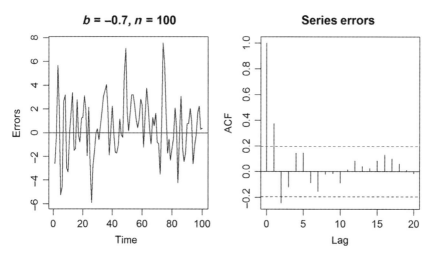

FIGURE 6.1   Case (i) with $b_1 = -0.7$ and white noise variance (σ_w^2) = 4.0.

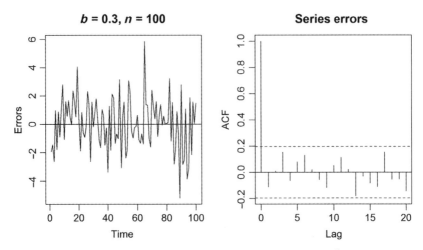

FIGURE 6.2   The second case with white noise variance $(\sigma_w^2) = 4.0$.

6.6   SIMULATED EXAMPLES OF THE MA(2) MODEL

Examples (white noise variance = 4.0): analogous to AR(2) models, MA(2) models can be thought of as two applications of an MA(1) model. The column "case" of Table 6.1 describes the nature of the b_1 values in each case.

For case (iii) (Figure 6.3), the sample values for this simulated data are $\hat{R}_1 = -0.602$, $\hat{R}_2 = 0.058$, $\hat{R}_3 = 0.077$, and $\hat{\sigma}_{MA}^2 = 2.98 \cdot \sigma_w^2$.

For case (iv) (Figure 6.4), the sample values for this simulated data are $\hat{R}_1 = -0.341$, $\hat{R}_2 = -0.133$, $\hat{R}_3 = -0.041$, and $\hat{\sigma}_{MA}^2 = 1.46\sigma_w^2$.

For case (v) (Figure 6.5), the sample values for this simulated data are $\hat{R}_1 = 0.341$, $\hat{R}_2 = -0.217$, $\hat{R}_3 = -0.127$, and $\hat{\sigma}_{MA}^2 = 1.28\sigma_w^2$.

For case (vi) (Figure 6.6), the sample values for this simulated data are $\hat{R}_1 = -0.715$, $\hat{R}_2 = 0.247$, $\hat{R}_3 = 0.001$, and $\hat{\sigma}_{MA}^2 = 2.63\sigma_w^2$.

In general, the sample values are close to the theoretical values found in Table 6.1.

6.7   AR(*m*) AND MA(*l*) MODEL *acf()* PLOTS

From both the calculations performed in Chapters 5 and 6 and the *acf()* plots, an often noted pattern can be observed in plots with known underlying model. That

TABLE 6.1   Examples of MA(2) Models

Case	b_1	b_2	R_1	R_2	R_3	σ_{MA}^2
(iii) +,+	1.5	−0.56	−0.657	0.157	0	$3.56\sigma_w^2$
(iv) large +,small −	0.5	0.24	−0.291	−0.184	0	$1.31\sigma_w^2$
(v) large −,small +	−0.5	0.24	0.291	−0.184	0	$1.31\sigma_w^2$
(vi) complex	1.2	−0.72	−0.698	0.243	0	$2.96\sigma_w^2$

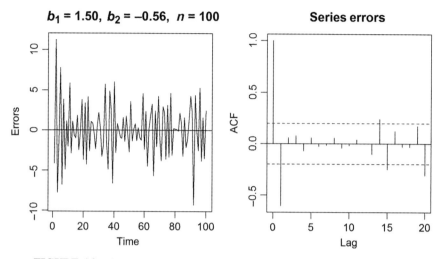

FIGURE 6.3   Case (iii). Equivalent to two iterations of MA(1) with positive *b*.

MA(*l*) models will show a number of spikes in the autocorrelation plot indicating the order. For example, ignoring the spike at lag zero of 1 (which always exists regardless of the data), the MA(1) model has just one spike at lag 1 (Figures 6.1 and 6.2), and the MA(2) model has just two spikes at lags 1 and 2 (Figures 6.3, 6.4, 6.5, and 6.6).

On the other hand, AR(*m*) models tend to show a slow decay and are somewhat less prone to have just a few sharp spikes (Figures 5.5, 5.6, and 5.7, and 5.9, 5.10, 5.11, and 5.12). Three comments are in order: (i) This pattern will continue for

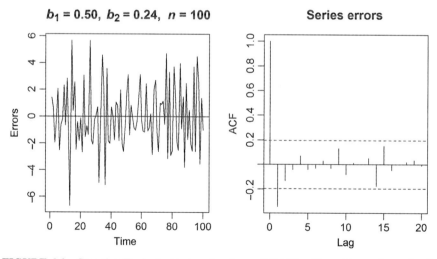

FIGURE 6.4   Case (iv). Equivalent to two iterations of MA(1) with positive and negative *b*.

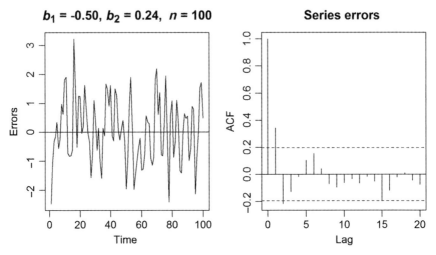

FIGURE 6.5   Case (v). Equivalent to two iterations of MA(1) with positive and negative b.

higher order MA(l) and AR(m) models; (ii) there is an a *pacf()* function, which will be discussed much later, that produces just the opposite pattern, a few sharp spikes for AR(m) models, indicating the order, and a slow decay for MA(l) models; and (iii) while these tendencies exist in the data and the plots, the plots are not easy to interpret in every case, especially when sample sizes are small.

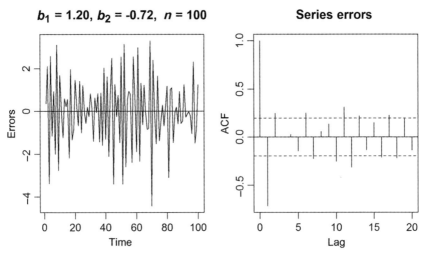

FIGURE 6.6   Case (vi). Equivalent to two iterations of MA(1) with complex conjugates for b.

EXERCISES

1. Derive all autocovariances and autocorrelations for the MA(1) model.

2. Derive all autocovariances and autocorrelations for the MA(2) model.

3. Give a rationale for MA(1) = AR(∞).

    (i) Show $\varepsilon_n = -b_1 \varepsilon_{n-1} - b_1^2 w_{n-2} + w_n$.

    (ii) Using the idea above, develop an MA(1) formula analogous to the AR(1) formula: $\varepsilon_n = a_1^k \varepsilon_{n-k} + \sum_{j=1}^{k} a_1^{j-1} w_{n-j+1}$.

    (iii) Use this formula to suggest an MA(1) process is equivalent to an AR(∞) process.

4. Solve for the values in the following table (with $\sigma_w^2 = 4.0$).

b_1	b_2	R_1	R_2	R_3	σ_{MA}^2
-0.6	0.3	??	??	??	??
0.1	-0.8				
-0.7	-0.2				
0.2	0.5	??			??

5. Simulate each series in the previous table with the specified white noise variance:

    (i) Verify b_1 and b_2 that meet the conditions for invertibility.

    (ii) Compute all values on the table.

    (iii) Simulate and plot the time series.

    (iv) Produce the *acf()* graph.

    (v) Verify R_1, R_2, R_3 and the variance for the series.

PART II

ANALYSIS OF PERIODIC DATA AND MODEL SELECTION

7

REVIEW OF TRANSCENDENTAL FUNCTIONS AND COMPLEX NUMBERS

7.1 BACKGROUND

A significant part of this book involves the fitting of periodic functions to periodic data. Many derivations in later chapters will also require understanding of the relationships between periodic functions and exponential functions. These are very powerful techniques, but they are based on some deep mathematics. Although the focus here is not on the proofs behind these methods, even the use of these methods, if any insight is desired, requires some very advanced algebraic and trigonometric manipulations. Besides, this is very cool mathematics.

One key formula associated with many of the manipulations performed throughout the rest of the book is due to Euler: $e^{ix} = \cos(x) + i \cdot \sin(x)$. A standard rationale for this formula, based on power series and complex numbers, will be given. Much of the mathematics in later chapters involves moving back and forth between the exponential and trigonometric forms of certain expressions. It is worth noting that a special case of this formula, $e^{i\pi} + 1 = 0$, is considered by many to be the most beautiful formula in mathematics.

Basic Data Analysis for Time Series with R, First Edition. DeWayne R. Derryberry.
© 2014 John Wiley & Sons, Inc. Published 2014 by John Wiley & Sons, Inc.

7.2   COMPLEX ARITHMETIC

7.2.1   The Number i

Many students who take statistics may have never been exposed to complex numbers at all. Complex numbers can be loosely considered linear combinations of real and imaginary numbers. The core of imaginary numbers is the quantity $i = \sqrt{-1}$. Any number of the form $a + bi$, where a and b are real, is a complex number. What follows may be a lot of new information, but it is just simple algebra.

$$i^2 = (\sqrt{-1})^2 = -1, \quad i^3 = -1 \cdot i = -i, \quad i^4 = (i^2)^2 = 1,$$

$$i^5 = i \cdot i^4 = i, \quad i^6 = i^2 \cdot i^4 = -1, \quad \text{etc.}$$

Sums, differences, products, and quotients of complex numbers are also complex numbers. For example $(3 - 2i) + (3 - 7i) = 6 - 9i$, $(3 - i) - (7 + 4i) = -4 + 3i$, and $(1 - 2i)(2 - i) = 2 - 4i - i + 2i^2 = -5i$. But what about $(3 - 2i)/(1 - 2i) = ?$

7.2.2   Complex Conjugates

The pair of complex numbers $a \pm bi$ are called complex conjugates and have been mentioned, without explanation, before. Complex conjugates are the key to simplifying many messy complex expressions. The first important property of complex conjugates is that their product is always real: $(a + bi)(a - bi) = a^2 - abi + abi - b^2i^2 = a^2 + b^2$.

This is the key idea required for the previous problem:

$$\frac{3 - 2i}{1 - 2i} = \frac{(3 - 2i)(1 + 2i)}{(1 - 2i)(1 + 2i)} = \frac{3 - 2i + 6i - 4i^2}{1^2 + 2^2} = \frac{7}{5} + \frac{4}{5}i.$$

Another notation that is often useful is $\overline{a + bi} = a - bi$ and $\overline{a - bi} = a + bi$, so that $\bar{z}$ is the conjugate of z. The complex conjugate of a number is always just the same number with the sign of the imaginary part reversed.

This is a powerful tool. For example, it is quite useful to know that

$$\overline{e^{ix}} = \cos(x) - i \cdot \sin(x).$$

7.2.3   The Magnitude of a Complex Number

An important way of thinking about complex numbers is that of a plane with a real axis and an imaginary axis, so the number $a + bi$ is a units to the right and b units above the origin in a two-dimensional plane (if a is negative, a units to the right is actually $|a|$ units to the left, so the statement is quite general).

On the real line the magnitude of a number a, $|a|$, is the distance from the origin. In complex mathematics this is also the magnitude, distance from the origin. The distance from the origin to the number $a + bi$ is then $\sqrt{a^2 + b^2}$ (just like any number in a plane).

Complex conjugates can be used to represent magnitude: Let $z = a + bi$ and use $|z|$ to represent magnitude (distance from the origin). Then $|z|^2 \neq (a + bi)^2$. Instead, $|z|^2 = z \cdot \bar{z} = a^2 + b^2$ and $|z| = \sqrt{a^2 + b^2}$.

7.3   SOME IMPORTANT SERIES

7.3.1   The Geometric and Some Transcendental Series

A number of the standard infinite series are widely useful in what follows.

The geometric series, often written in the form $\sum_{j=0}^{m} ar^k = a + ar + ar^2 \ldots$, will be used in at least one rather important derivation.

The summation can be found using a standard trick which is explained as follows. $S = a + ar + ar^2 + \ldots ar^m$ now creates a second series by multiplying the first by r, $rS = ar + ar^2 + ar^3 + \ldots ar^{m+1}$. Taking the difference yields $S - rS = a - ar^{m+1}$, which produces $S = a(1 - r^{m+1})/(1 - r)$. Furthermore, when $|r| < 1$, $\lim_{m \to \infty} S = a/(1 - r)$.

7.3.2   A Rationale for Euler's Formula

Three other important series found in every textbook on second semester calculus are

$$\sin(x) = \sum_{k=1}^{\infty} \frac{(-1)^{k+1} x^{2k-1}}{(2k-1)!}, \quad \cos(x) = \sum_{k=0}^{\infty} \frac{(-1)^k x^{2k}}{(2k)!}, \quad \text{and } \exp(x) = \sum_{k=0}^{\infty} \frac{x^k}{k!}.$$

Given these three series, it is easy to derive Euler's formula, assuming it is valid to casually re-group the terms in an infinite series. Because some of the technical details have been ignored, this is a rationale and not a proof, although the result is correct.

$$\exp(ix) = \sum_{k=0}^{\infty} (ix)^k = 1 + ix + \frac{i^2 x^2}{2!} + \frac{i^3 x^3}{3!} + \frac{i^4 x^4}{4!} + \frac{i^5 x^5}{5!} + \cdots$$

$$= \left\{ 1 - \frac{x^2}{2!} + \frac{x^4}{4!} - \frac{x^6}{6!} + \cdots \right\} + i \left\{ x - \frac{x^3}{3!} + \frac{x^5}{5!} - \frac{x^7}{7!} + \cdots \right\}$$

$$= \cos(x) + i \cdot \sin(x).$$

7.4 USEFUL FACTS ABOUT PERIODIC TRANSCENDENTAL FUNCTIONS

Manipulating sine and cosine functions will become quite routine. Some frequently used facts include

(i) $\sin(-x) = -\sin(x)$

(ii) $\cos(-x) = \cos(x)$

(iii) $\sin(x + y) = \sin(x)\cos(y) + \cos(x)\sin(y)$

(iv) $\cos(x + y) = \cos(x)\cos(y) - \sin(x)\sin(y)$

(v) $\overline{\exp(iu)} = \cos(u) - i \cdot \sin(u) = \exp(-iu)$

(vi) $\exp(iu) \cdot \overline{\exp(iu)} = \cos^2(u) + \sin^2(u) = 1$. In other words, $|\exp(iu)| = 1$.

EXERCISES

1. Simplify $\dfrac{2 - 3i}{1 - 2i}$.

2. Simplify $\dfrac{c + di}{a - bi}$.

3. Compute the sum: $S = 2 + 1 + 0.5 + 0.25 + \ldots$

4. Evaluate $\exp(iu) + \exp(-iu) = ??$

5. Show $\cos(x) = (\exp[ix] + \exp[-ix])/2$ and $\sin(x) = (\exp[ix] - \exp[-ix])/2$.

6. (i) Show $\sum_{s=0}^{n-1} \exp(2\pi sfi) = \dfrac{\exp(2\pi nfi) - 1}{\exp(2\pi fi) - 1}$.

(ii) Evaluate the sum at $f = 0$, $1/3$, 0.5, 1, when $n = 12$.

7. (i) Show $\sum_{s=0}^{n-1} \cos(2\pi sf) = \cos(\pi f[n - 1])\dfrac{\sin(\pi fn)}{\sin(\pi f)}$ (hint: use induction).

(ii) Evaluate the sum at $f = 0$, 0.5, $2/3$, 1, when $n = 24$.

8. Show $\sin(2x) = 2 \cdot \sin(x) \cdot \cos(x)$ and $\cos(2x) = \cos^2(x) - \sin^2(x)$.

9. Use the series representations to show:

(a) $\dfrac{d}{dx}\exp(x) = \exp(x)$.

(b) $\dfrac{d}{dx}\sin(x) = \cos(x)$.

(c) $\dfrac{d}{dx}\cos(x) = -\sin(x)$.

8

THE POWER SPECTRUM AND THE PERIODOGRAM

8.1 INTRODUCTION

This chapter is concerned with estimation of a function, $p(f)$, that uses all of the covariances, C_k, to form a periodic function for all f, thought of as frequencies. Although it is hard to motivate this function and it seems to be pulled out of thin air, it will be important throughout the rest of the book. The (power) spectrum is the true function, $p(f)$, while the periodogram is $\hat{p}(f)$ and is estimated from the data. While $p(f)$ is based on all covariances ($k = 0,1,2,3,\ldots$) and exists for all frequencies, f, the analogous sample estimate $\hat{p}(f)$, must be limited by sample size, and uses the sample covariances $k = 0, 1, \ldots n - 1$ and can only be computed at $f_j = j/n$, where $j = 1, 2 \ldots n/2$. Because $p(f)$ is periodic, it need only be specified in the interval $0 \leq f \leq 1$ or $-0.5 \leq f \leq 0.5$, furthermore the condition $C_k = C_{-k}$ renders half of this information redundant, so it is sufficient to define the function in the interval $0 \leq f \leq 0.5$.

The spectrum measures the intensity of periodic patterns of different frequencies. The periodic ripples could be transient and fading (as with AR(m) and MA(l) models) or persistent (as with periodic signals). Plots of $\log_e[\hat{p}(f)]$ versus the frequencies provide the second useful plot unique to time series and the rationale for the R function *spec.pgram()*. For example, sales data measured monthly will have a maximum of the spectrum, hopefully reflected in its estimate, the periodogram, at $f_j = 1/12$ (because

Basic Data Analysis for Time Series with R, First Edition. DeWayne R. Derryberry.
© 2014 John Wiley & Sons, Inc. Published 2014 by John Wiley & Sons, Inc.

high and low sales will occur at the same time annually). Some might fancifully think of monthly sales data as having a wavelength of 12.

The various AR(m) and MA(l) models have their own unique graphs, and periodic signals stand out in the periodogram and spectrum.

8.2   A DEFINITION AND A SIMPLIFIED FORM FOR $p(f)$

The spectrum is defined as $p(f) = \sum_{k=-\infty}^{\infty} C_k \exp(-2\pi kfi)$ and need only to be considered in the interval $0 < f < 1$ or $-\frac{1}{2} < f < \frac{1}{2}$, but it is easy to see $\sum_{k=-\infty}^{\infty} C_k \exp(-2\pi kfi) = \sum_{k=-\infty}^{\infty} C_k[\cos(2\pi kf) + i \cdot \sin(2\pi kf)]$, by Euler's formula, of course. But $\sum_{k=-\infty}^{\infty} iC_k \cdot \sin(2\pi kf) = 0$ (Exercise 1). So $p(f) = \sum_{k=-\infty}^{\infty} C_k \cdot \cos(2\pi kf) = C_0 \cos(0) + \sum_{k=1}^{\infty} C_k \cos(2\pi kf) + \sum_{k=-1}^{-\infty} C_k \cos(2\pi kf) = C_0 + 2\sum_{k=1}^{\infty} C_k \cos(2\pi kf)$.

8.3   INVERTING $p(f)$ TO RECOVER THE C_k VALUES

Surprisingly, it is also the case that $C_k = \int_{-1/2}^{1/2} p(f) \exp(2\pi kfi) df$, for $k = 0, 1, 2, 3, \ldots$

In other words, there is a one-to-one relationship between the complete auto-correlation function and the power spectrum. Showing this will involve a bit more work.

Note: In what follows, the order of integration and summation is exchanged without much apparent thought. In advanced calculus, it can be shown that this can be problematic. Nevertheless, it is well known that those problems do not arise in this case:

$$C_k = \int_{-\frac{1}{2}}^{\frac{1}{2}} p(f) \exp(2\pi kfi) df = \int_{-\frac{1}{2}}^{\frac{1}{2}} p(f) \cos(2\pi kf) df + i \int_{-1/2}^{1/2} p(f) \sin(2\pi kf) df.$$

The next step is to use the cosine representation of $p(f)$ derived above. This will produce a product of cosines in the first integral and a product of sines and cosines in the second integral. The second integral will vanish (Exercise 2), while the first will simplify quite miraculously (for those new to Fourier series). This little section might inspire someone to learn more about Fourier series, generalized inner products, and Orthogonal functions. Focusing on the part that does not vanish:

$$?? = \int_{-1/2}^{1/2} p(f) \cos(2\pi kf) df = \int_{-1/2}^{1/2} \left\{ C_0 + 2\sum_{j=1}^{\infty} C_j \cos(2\pi jf) \right\} \cos(2\pi kf) df$$

$$= \int_{-1/2}^{1/2} C_0 \cos(2\pi kf) df + 2\sum_{j=1}^{\infty} \int_{-1/2}^{1/2} C_j \cos(2\pi jf) \cos(2\pi kf) df.$$

Suppose $k = 0$ [recall $\sin(m\pi) = 0$ for any integer m, to be used near the end].

$$?? = \int_{-\frac{1}{2}}^{\frac{1}{2}} C_0 \cos(0) df + 2 \sum_{j=1}^{\infty} \int_{-\frac{1}{2}}^{\frac{1}{2}} C_j \cos(2\pi jf) \cos(0) df$$

$$= \int_{-1/2}^{1/2} C_0 df + 2 \sum_{j=1}^{\infty} C_j \int_{-1/2}^{1/2} \cos(2\pi jf) df = C_0 + 2 \sum_{j=1}^{\infty} C_j \frac{-\sin(2\pi jf)}{2\pi j} \bigg]_{-1/2}^{1/2}$$

$$= C_0$$

Suppose $k \neq 0$.

In general, $\int_{-1/2}^{1/2} \cos(2\pi mf) df = 0$ for any integer m (as shown in the integral immediately above.). So

$$?? = C_0 \int_{-1/2}^{1/2} \cos(2\pi kf) df + 2 \sum_{j=1}^{\infty} C_j \int_{-1/2}^{1/2} \cos(2\pi jf) \cos(2\pi kf) df$$

$$= 0 + 2 \sum_{j=1}^{\infty} \cos(2\pi jf) \cos(2\pi kf) df.$$

But $2 \cdot \cos(2\pi jf) \cos(2\pi kf) = \cos(2\pi[j + k]f) + \cos(2\pi[j - k]f)$ [Exercise 3]:

$$?? = 2 \sum_{j=1}^{\infty} \frac{1}{2} C_j \int_{-1/2}^{1/2} \{\cos(2\pi[j - k]f) + \cos(2\pi[j + k]f)\} df$$

Now consider $j = k$ and $j \neq k$, term by term in the summation.

If $j = k$, the term becomes $\frac{1}{2} C_k \int_{-1/2}^{1/2} \{\cos(0) + \cos(4\pi kf)\} df = \frac{1}{2} C_k$.

If $j \neq k$, then, where m and r are integers different from zero, $\frac{1}{2} C_j \int_{-1/2}^{1/2} \{\cos(2\pi mf) + \cos(2\pi rf)\} df = 0$, so $?? = 2 \sum_{j=1}^{\infty} \frac{1}{2} C_j \int_{-1/2}^{1/2} \{\cos(2\pi[j - k]f) + \cos(2\pi[j + k]f)\} df = C_k$.

Hence, it is verified that the expression given produces C_k, for $k = 0, 1, 2, \ldots$ At some point a mathematician might interject that there is no need for both an autocorrelation function and a power spectrum. Both are just different presentations of the same information. Nevertheless, there are two reasons for studying both. The first is that different presentations of the same information highlight different things; the second, and more important, is that statisticians work with others and those others might have a strong preference (determined by their field, training, and journals) for one function or the other.

8.4   THE POWER SPECTRUM FOR SOME FAMILIAR MODELS

8.4.1   White Noise

White noise has $C_k = 0$ for all $k > 0$. The power spectrum is simply the variance:

$$p(f) = C_0 = \sigma_w^2.$$

Figure 8.1 represents the periodogram, on the logarithmic scale, when there is no pattern (A method for computation of the periodogram will be given in Section 8.5.) The spectrum, which the periodogram is estimating, is $\log_e[p(f)] = \log_e(9)$, which is the horizontal line on the periodogram. Since $acf()$ plots and periodograms contain the same information in different formats, plotting them together will allow the reader to think of them together and decide for themselves which better conveys the underlying information.

8.4.2   The Spectrum for AR(1) Models

The power spectrum for AR(1), $p(f) = \sum_{k=-\infty}^{\infty} C_k \exp(-2\pi kfi)$, is moderately challenging to derive [recall $C_k = a^{|k|} C_0$ and $C_0 = \frac{\sigma_w^2}{1-a^2}$ for AR(1)].

Beginning with the definition in terms of cosines and covariances, the series is divided into two convergence geometric series:

$$\sum_{k=-\infty}^{\infty} C_k \exp(-2\pi kfi) = \sum_{k=-\infty}^{\infty} a^{|k|} e^{-2\pi kfi} = C_0 + \sum_{k=1}^{\infty}(ae^{-2\pi fi})^k + \sum_{j=1}^{\infty}(ae^{2\pi fi})^j.$$

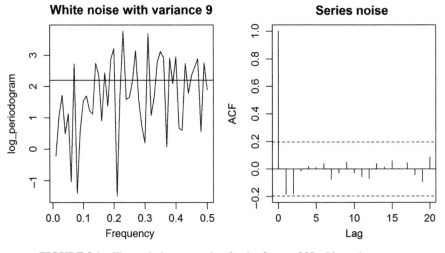

FIGURE 8.1   The periodogram and $acf()$ plot for $n = 200$ white noise errors.

It is trivial to show $\sum_{k=1}^{\infty}(ae^{-2\pi fi})^k = \frac{ae^{-2\pi fi}}{1-ae^{-2\pi fi}}$ and $\sum_{j=1}^{\infty}(ae^{2\pi fi})^j = \frac{ae^{2\pi fi}}{1-ae^{2\pi fi}}$. (How is it known these series are convergent?)

The hard part, which is all just tedious algebra, follows:

$$p(f) = \frac{\sigma_w^2}{1-a^2}\left\{1+\frac{ae^{-2\pi fi}}{1-ae^{-2\pi fi}}+\frac{ae^{2\pi fi}}{1-ae^{2\pi fi}}\right\},$$

forming a common denominator, produces

$$\frac{\sigma_w^2}{1-a^2}\left\{\frac{1-a^2}{(1-ae^{-2\pi fi})(1-ae^{2\pi fi})}\right\} = \frac{\sigma_w^2}{1-2a\cdot\cos(2\pi f)+a^2} = \frac{\sigma_w^2}{|1-e^{-2\pi fi}|^2}.$$

The bottom left expression makes the computation of the spectrum easy, while the bottom right expression makes the general pattern more transparent [as can be seen with the analogous AR(2) expressions to follow shortly].

The $acf()$ plot and spectrum, on the logarithmic scale, when $a = 0.9$, $\sigma_w^2 = 9$ and -0.5 are below ($n = 200$).

For an AR(1) model with positive a_1, Figure 8.2, the highest correlations are at the most nearby observations, so the spectrum peaks are at $f = 0$. For negative a_1, Figure 8.3, the minimum is at $f = 0$ and the maximum is at $f = 0.5$. The magnitude of a_1 determines the steepness of the curve.

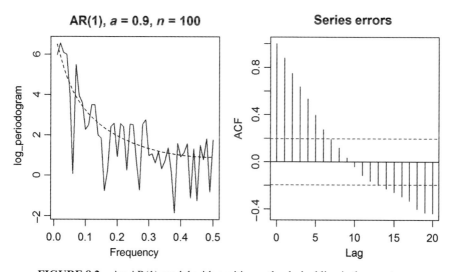

FIGURE 8.2 An AR(1) model with positive a, the dashed line is the spectrum.

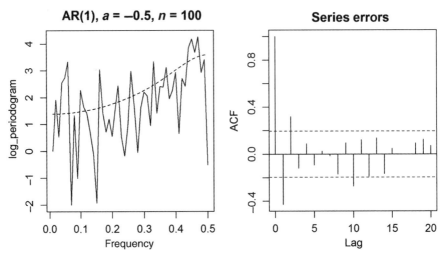

FIGURE 8.3   An AR(1) model with negative a, the dashed line is the spectrum.

8.4.3   The Spectrum for AR(2) Models

The spectrum for the AR(2) model is (derivation will require much more advanced ideas) $p(f) = \sigma_w^2/[1 - 2a_1 \cos(2\pi f) - 2a_2 \cos(4\pi f) + 2a_1 a_2 \cos(2\pi f) + a_1^2 + a_2^2] = \sigma_w^2/|1 - a_1 e^{-2\pi f i} - a_2 e^{2\pi f i}|^2$.

As previously, four typical examples will be given (both a's positive, mixed signs, and complex conjugates).

The first example, Figure 8.4, is not much different from an AR(1) model with positive a_1.

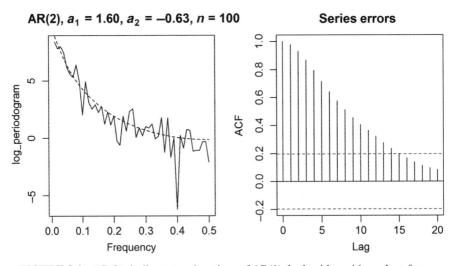

FIGURE 8.4   AR(2) similar to two iterations of AR(1), both with positive values for a_1.

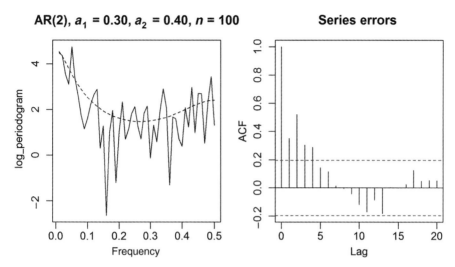

FIGURE 8.5   AR(2) similar to two iterations of an AR(1) process with one positive and one negative *a*, the positive value is larger in magnitude.

Figures 8.5 and 8.6 show AR(2) models when they are equivalent to two iterations of an AR(1) model with a_1 alternating in sign. Notice that the differences between these two models are more pronounced in the *acf()* plots.

The AR(2) model is "equivalent" to two iterations of an AR(1) model; but with complex conjugate values for a_1, it produces "pseudoperiodic" behavior (Box et al., 2008, pp. 62–65). This is displayed in Figure 8.7.

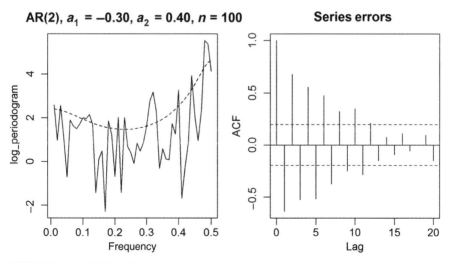

FIGURE 8.6   AR(2) similar to two iterations of an AR(1) process with one positive and one negative *a*, the negative value larger in magnitude.

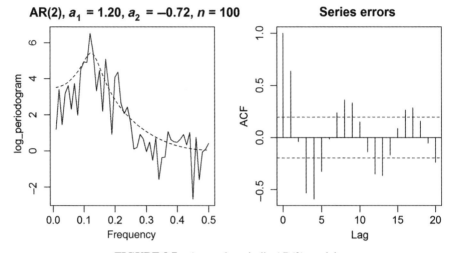

FIGURE 8.7   A pseudoperiodic AR(2) model.

This model highlights that AR(m) and MA(l) patterns in noise can be hard to distinguish from periodic patterns in the signal.

8.5   THE PERIODOGRAM, A CLOSER LOOK

8.5.1   Why is the Periodogram Useful?

The periodogram is ideal for identifying periodicity in data and estimating the frequency of the period. Consider a very simple periodic pattern with no noise producing Figure 8.8. Periodic functions will be discussed in more detail in Chapter 10.

```
n <- 200
x <- c(1:n)
y <- 5 + 4*cos(2*pi*x/10+.25)     # a periodic signal with period 10
plot(x,y)
lines(x,y)
```

A close inspection of the periodogram (right panel of Figure 8.8) will show that one strong frequency is at 1/10, indicating the correct period of 10.

If a little noise is added to y: err <- rnorm(n,0,0.25), y_err <- y + err, the data appears like in Figure 8.9 and does not appear much different from Figure 8.8, [nor are the *acf( )* plots much effected] but the periodograms are very different.

8.5.2   Some Naïve Code for a Periodogram

Computation of the periodogram, which is the sample estimate of the power spectrum, is quite straightforward. Since $p(f) = C_0 + 2 \sum_{k=1}^{\infty} C_k \cos(2\pi k f)$, the obvious

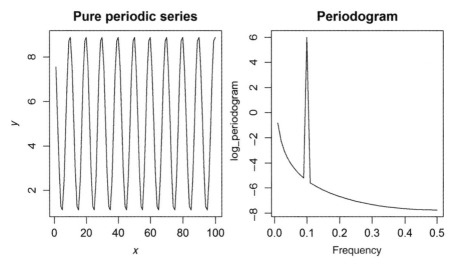

FIGURE 8.8   Pure periodic behavior with no noise.

approach for the sample estimate is $\hat{p}(f) = \hat{C}_0 + 2 \sum_{k=1}^{n-1} \hat{C}_k \cos(2\pi kf)$. While $p(f)$ is defined for all $0 < f < 0.5$ [it exists outside this interval, but because of symmetry and periodicity, it only needs to be known in this interval], $\hat{p}(f)$ can only be computed as certain discrete values in that range, $f = j/n$, where $j = 1 \ldots n/2$ when n is even and $j = 1 \ldots (n-1)/2$ when n is odd. Of course, all $\hat{C}_k$ can be recovered from $acf()$ in R, so computation of the periodogram is quite straightforward.

Strategy for computation of a periodogram: the approach involves two "for" loops. The outer loop passes through the frequencies and the inner loop sums over all

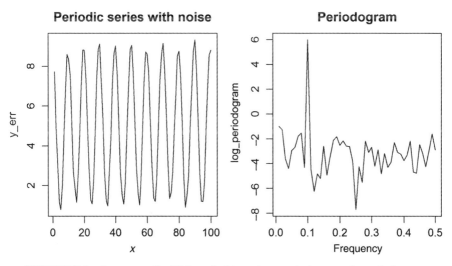

FIGURE 8.9   A very small addition of white noise greatly impacts the periodogram.

$n - 1$ autocovariances, summing products of covariances and cosines [using $\hat{p}(f) = \hat{C}_0 + 2\sum_{k=1}^{n-1} \hat{C}_k \cos(2\pi f)$].

Get estimated autocorrelations ($\hat{C}_k$) and variance from *acf()*

Set: $\hat{p}(f)$ to zero as a vector of length 1:floor(n/2) *floor()* rounds down to an integer

Add $\hat{C}_0$ to every entry in the $\hat{p}(f)$ vector

for each frequency $f_j = j/n$,

for $k = 1$ to $n - 1$, add $2\hat{C}_k \cos(2\pi kf_j)$ to $\hat{p}(f_j)$

end

end

Naïve R code for the periodogram is

```
y <- scan(data...)      # read in a time series to analyze
n <- length(y)              #determine the length of the data
# use the acf() function to recover all autocovariances.
w <- acf(y, lag.max = n, type = c("covariance"))
# while the covariances range from 0 to n-1, the vector that stores them in acf()
# ranges from 1 to n, in this case that would be  w$acf[1:n]
C <- w$acf[1:n]
C_0 <- w$acf[1]
m <- floor(n/2)
freq <- c(1:m)/n
# p_hat is the periodgram in this code. It is floor(n/2) long.
p_hat <- rep(C_0,m)
 for(j in 1:m)              # pass through the frequencies
{ f <- freq[j]
      for(k in 1:(n-1))                  #sum over the covariances and cosines
     { p_hat[j] <- p_hat[j] + 2*C[k+1]*cos(2*pi*k*f)}
}
```

8.5.3   An Example—The Sunspot Data

The sunspot data (Figure 3.2), which will be analyzed in detail in Chapter 17, appears to be periodic with an unknown period. A periodogram can be quite helpful in exploring this data. (It makes sense to transform the data, but this analysis is quite preliminary and quite brief.)

The *acf()* plot in Figure 8.10 differs from some others in that (i) the covariances are computed instead of the default correlations (for variety) and (ii) covariances at all lags have been computed. When examining an *acf()* plot for periodicity, it is valuable to plot all lags to see if the period persists or decays (as is the case here). It will be determined in Chapter 17 that the sunspot data does display relatively periodic

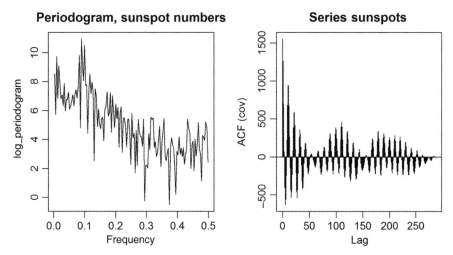

FIGURE 8.10   The sunspot numbers seem to be periodic, but the period is unclear.

behavior, but that the characteristics change over time. However, it would already be a good guess that the period must be about 10 years [the periodogram has a spike at about 1/10 and the *acf()* plot has about 10 peaks between 0 and 100 years] even if it varies a bit from period to period.

In order to explore this kind of data further, methods introduced in Chapter 10 are required to fit periodic functions to data.

8.6   THE FUNCTION *SPEC.PGRAM()* IN R

It is important to know how a periodogram is constructed, but there is already a function in R that computes the periodogram. Except for a few exercises at the end of this chapter where the student is specifically asked to compute the periodogram by hand, it will always be permitted to use *spec.pgram()*, the R function.

The default code, for the purposes of this book, to get a periodogram in R will be z <- spec.pgram(y, fast = FALSE, taper = 0.0), which produces a periodogram for data "y" identical to the naïve code created above.

The code fast = FALSE and taper = 0.0 should be explained. The naïve code developed earlier in this chapter is not the default used by R, but it will be the default for this book, so these codes override default settings in R.

Why fast = FALSE? The Fast Fourier Transform (FFT) is a technique of speeding up the algorithm that computes the periodogram. While this algorithm substantially speeds the process of computing a periodogram, it does alter the frequencies and has other minor side effects. For the data sets in this book, and with modern computers, the naïve code runs essentially instantly. The command fast = FALSE changes the default from the FFT to the direct method in the naïve code. (When dealing with

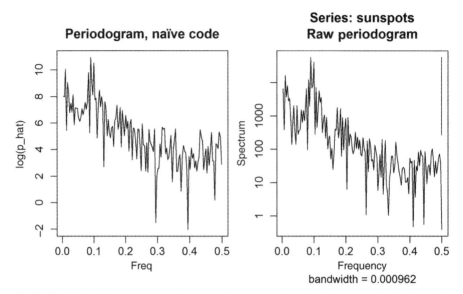

FIGURE 8.11 The sunspot number periodogram produced with naïve code (left) and R defaults (right).

huge data sets, the function *spec.pgram( )* may run slowly, in that case the default with fast = TRUE will produce enormous time savings.)

The command taper = 0.0 overrides the default taper = 0.1, a procedure that tapers the ends of the series. Tapering is related to leakage (Bloomfield, 2002) and is not discussed in detail here. A number of modifications intent of improving on the periodogram will be enlisted, but the motivations will relate to the bias–variance tradeoff discussed in detail in the next chapter. As always, help(spec.pgram) is quite helpful.

From Figure 8.11, it is obvious that whether the R default values or the values fast = FALSE, taper = 0.0 (naïve code) are used; the sunspot data has a very similar periodogram. Some apologies may be in order. Many will be greatly offended by the dismissal of the FFT (the FFT is widely considered one of the top 10 algorithms of the twentieth century), leakage, windows, and tapering, as discussed in traditional texts on Fourier series (Bloomfield). In fact, the approach here is simply to begin with the periodogram naively defined and modify that periodogram based on more mainstream data analysis ideas. (This may be the one place where writing for data analysts, rather than a traditional time series audience, leads the book down a different path.)

This book is the beginning of a journey, not the end, and Bloomfield's book is still worth reading in its entirety—a task made easier by beginning with this book.

Returning to the command z <- spec.pgram(y, fast = FALSE, taper = 0.0), then names(z)

Produces the output: "freq" "spec" "coh" "phase" "kernel" "df"
"bandwidth" "n.used" "orig.n" "series" "snames" "method"
"taper" "pad" "detrend" "demean".

The frequencies are z$freq and the values for the periodogram are z$spec, so plot(z$freq,log(z$spec)) produces a crude plot numerically identical to the plot produced by *spec.pgram()*.

EXERCISES

The first three exercises relate to the simplification of the spectrum to an expression in autocovariances and cosines.

1. Show $\sum_{k=-\infty}^{\infty} i \cdot C_k \sin(2\pi kf) = 0$. Hints: (i) recall $C_k = C_{-k}$ and (ii) what is the value of $\sin(b) + \sin(-b)$?

2. Show $\int_{-1/2}^{1/2} p(f) \sin(2\pi kf) df = 0$.
    (i) Show $\sin(\alpha + \beta) + \sin(\alpha - \beta) = 2 \sin(\alpha) \cos(\beta)$.
    (ii) $\int_{-1/2}^{1/2} p(f) \sin(2\pi kf) df = \int_{-1/2}^{1/2} \{C_o + 2 \sum_{j=1}^{\infty} C_j \cos(2\pi jf)\} \sin(2\pi kf) df$, why?
    (iii) Multiply out the terms of the initial integral and simplify using (i) and (ii).

3. Show $2 \cdot \cos(2\pi jf) \cos(2\pi kf) = \cos(2\pi[j + k]f) + \cos(2\pi[j - k]f)$.

4. Justify the claim that the following series related to the derivation of the AR(1) spectrum are convergent:

$$\sum_{k=1}^{\infty} (ae^{-2\pi fi})^k = \frac{ae^{-2\pi fi}}{1 - ae^{-2\pi fi}} \quad \text{and} \quad \sum_{j=1}^{\infty} (ae^{2\pi fi})^j = \frac{ae^{2\pi fi}}{1 - ae^{2\pi fi}}.$$

5. Show $|1 - ae^{-2\pi fi}|^2 = 1 - 2a \cdot \cos(2\pi f) + a^2$ and $|1 - a_1 e^{-2\pi fi} - a_2 e^{-4\pi fi}|^2 = 1 - 2a_1 \cos(2\pi f) - 2a_2 \cos(4\pi f) + 2a_1 a_2 \cos(2\pi f) + a_1^2 + a_2^2$ [this is just the real expansion of the denominator of the AR(2) power spectrum].

6. The periodogram of an AR(2) model…
    (i) Find the closed form formula for the maximum or minimum of

    $$h(f) = 1 - 2a_1 \cos(2\pi f) - 2a_2 \cos(4\pi f) + 2a_1 a_2 \cos(2\pi f) + a_1^2 + a_2^2 \text{ in}$$
    $$0 < f < 0.5.$$

    (ii) Compute the second derivative of $h(f)$.
    (iii) Use this information to discuss the graphs of $\log_e(p[f])$ for AR(2) models in Figures 8.4, 8.5, 8.6, and 8.7.

7. Plots for MA(1) and MA(2) similar to Figures 8.2, 8.3, 8.4, 8.5, 8.6, and 8.7.
    (i) Find $p(f)$ for the MA(1) and MA(2) models.
    (ii) Simulate 300 MA(1) observations and plot the periodogram ($\log_e$ scale) and the true model (derived above) both versus frequency. You can choose your

own parameter, but state what it is, and verify that it satisfies the invertibility condition.

   (iii) Simulate 300 MA(2) observations and plot the periodogram ($\log_e$ scale) and the true model (derived above) both versus frequency. You can choose your own parameters, but state what they are, and verify that they satisfy the invertibility conditions.

8. Produce a periodogram for the lynx data [From Exercise 1 (d), Chapter 3] and estimate the primary period (eyeball it). Use both the naïve code and R to produce the periodogram and verify that both methods are getting the same results.

9. Compute the *acf( )* plot and the periodogram for the mean maximum temperature in Melbourne [Exercise 1 (a), Chapter 3]. Identify (and/or reason to) the period for this data and explain what characteristic of each plot verifies this period.

10. (i) Simulate 400 AR(2) with $a_1 = 1.559$ and $a_2 = -0.81$ and plot the periodogram ($\log_e$ scale) and the true model both versus frequency.

   (ii) Using Exercise 6, compute where the maximum of $p(f)$ should occur [and that it is indeed a maximum] and that the graph matches this computation.

9

SMOOTHERS, THE BIAS-VARIANCE TRADEOFF, AND THE SMOOTHED PERIODOGRAM

9.1 WHY IS SMOOTHING REQUIRED?

It is obvious that $E(\hat{p}[f]) = E(\hat{C}_0) + 2 \sum_{k=1}^{n-1} E(\hat{C}_k) \cos(2\pi k f) = p(f)$. In other words, the expected value of the periodogram is equal to the spectrum. However, the periodogram is not a consistent estimator of the spectrum. No matter what sample size is chosen, the periodogram does not converge to the spectrum (Kitagawa, 2010, p. 38).

The reason is quite intuitive. All the values of $\hat{C}_k$ for $k = 1 \ldots n-1$ are required for the estimation of $\hat{p}(f)$ at any frequency f. When k is large (close to n), only a few sample values are used to estimate this covariance, so this value cannot be estimated consistently. Since the periodogram is a linear combination of estimates that are not consistent, it is a bit optimistic to hope that the result would be consistent. However, when sample size increases, there are always ways of averaging to get consistent estimators. See Box et al. (2008, pp. 41–42) for a slightly different intuitive explanation for the inconsistency of the periodogram.

9.2 SMOOTHING, BIAS, AND VARIANCE

So, can the periodogram be used to construct a consistent estimate of the spectrum?

The periodogram can be viewed as a collection of data points falling at random above or below the true model, at fixed frequencies.

The line in Figure 9.1 is the spectrum; the points are the periodogram ($\log_e$ scale).

As more observations are gathered on a fixed range of the x-axis (in our case $0 < f < 0.5$), more information about the true function over that range is collected.

Basic Data Analysis for Time Series with R, First Edition. DeWayne R. Derryberry.
© 2014 John Wiley & Sons, Inc. Published 2014 by John Wiley & Sons, Inc.

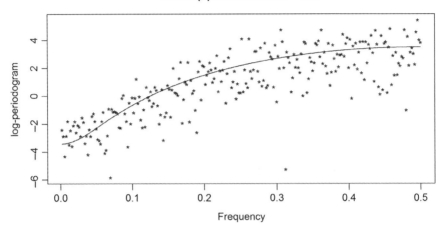

FIGURE 9.1   An MA(2) model with $b_1 = 1, 4, b_2 = -0.49$.

One common approach is to apply a local smoother to estimate the underlying true function.

Suppose the underlying true function can be assumed to be continuous and relatively smooth (without wild oscillations). Such a function can be viewed as locally approximately polynomial. Based on Taylor series ideas, such functions can be well approximated by polynomials of lower and lower order if the range of the x-axis becomes smaller and smaller. In this case, if we partition the x-axis into small segments, each one will contain points close to both local polynomial approximation and the true function value. As sample size increases, more data will fill each partition and the estimate within each partition will be reduced in variance.

All the smoothers considered have the property that they are exact for some order of polynomial functions. Reducing the width of each segment of the partition reduces the bias in this estimation procedure because the true function and a polynomial approximation become more alike. Increasing the number of observations in each segment of the partition reduces variance. Any approach that uses a "good" local smoother and constructs partitions so that increased sample sizes both reduce the width of the segments of the partitions and, at the same time, increase the number of data points within each segment of the partition can be used to smooth the periodogram to produce a consistent estimator of the spectrum.

9.3   SMOOTHERS USED IN R

Two approaches to smoothing the periodogram in R are useful. One approach uses smoothers that appear generally in R and can be used whenever smoothing is a useful idea, and the smoothers *lowess()* and *smooth.spline()* are presented (although other smoothers are available). The other approach uses a smoother built into *spec.pgram()*.

In all three cases, the approach will be the same. Local smoothers select points locally to average the estimation of the true function in that region of the x-axis. A strategy for smoothing is chosen so that, as the sample size increases, more points are included in each local estimate (variance reduction), yet each local estimate is over a smaller region of the x-axis (bias reduction). An obvious implementation of this approach is to include approximately $c\sqrt{n}$ observations in each local estimate, where c is a constant.

9.3.1 The R Function *Lowess()*

9.3.1.1 Background   The lowess smoother in R is a weighted local linear regression (lowess is a newer program that has more features and can fit higher order polynomials but does not produce a measure of the complexity of the fit, which will be required for later applications). Given a vector of x and a vector of y values, *lowess()* estimates the true function (at each value of x) by choosing some proportion of the available observations (this can be controlled by the user) near that x and fitting a polynomial (the order can be chosen by the user) using least squares regression with the observations closest to the x value being estimated having the most weight. *Lowess()* does this for every value of x and a line connecting these estimates is the estimate of the true function. When smoothing a periodogram, the x values will typically be the frequencies and the y values will typically be the log of the periodogram. [Do not be afraid to use help(lowess) to learn more].

The key user-controlled value is "f," the fraction of the data to include in each local estimate. Based on what has already been asserted, good results (a consistent estimator of the spectrum) can be obtained by smoothing the periodogram with $\approx \alpha/\sqrt{n}$, where α might vary, although it will usually be 1.0. This is not to say that this is an optimal smoothing, but that, as sample size increases, this smoothed estimate will converge to the underlying function producing the data.

Hopefully, there is no confusion, f is widely used for frequency in time series, while "f" is used for the fraction of the data to include in a local smoother in *lowess()*. The author is in no position to change either of these conventions.

9.3.1.2 An Example   To show how the procedure works, a large data set is simulated with a known true underlying function (Figure 9.2). The smoother mimics the true function well when $n = 2000$. The true model is an AR(2) model with, $a_1 = 0.5, a_2 = -0.25$, and a standard deviation of 3.

The data is simulated (it is assumed it is now obvious how to simulate such data) and named y; the command z <- spec.pgram(y, fast = FALSE, taper = 0.0) is used and

```
plot(z$freq,log(z$spec),  pch = ". ",xlab = "Frequency", ylab = "log-periodogram")
# true underlying true function
log_pt <- 2*log(3) -log(1 - 2*a1*(1-a2)*cos(2*pi*z$freq)-2*a2*cos(4*pi*z$freq)
+ a1^2 + a2^2)
lines(z$freq, log_pt)          # plot the true AR(2) function on the log scale
# implement lowess(), with a reasonable partition of the data
n <- length(z)
```

```
#smoothing the periodogram on the log scale
smoothfit <- lowess(z$freq, log(z$spec), f = 1/sqrt(n))
lines(smoothfit$x,smoothfit$y, lty = 2) # plot the resulting line
title("Spectrum and lowess() smoother[—]").
```

[Explore: It would be insightful to employ the command names(smoothfit) here].

The *lowess()* smoother underestimates, on average, but is doing a good job of producing a smoothed curve close to the true underlying curve. The points themselves, as an estimate of the true model, are not very useful without this smoothing. It is also possible to smooth the data on the "spec" scale but the data is more symmetric on the log scale and, more importantly, the smoother could produce negative estimates on the "spec" scale which would not allow for conversion to the log scale.

An important feature, which will be useful later, is the relationship $p = 2(1 + 1/f)$. In this case, this is $p = 2(1 + \sqrt{2000}) \approx 91.44$. This is the estimated complexity, number of free parameters, of the smoothed curve fitting through the data. Later, it will be important to compare different potential fits to the data by comparing their fit to the data relative to their complexity.

9.3.2   The R Function *Smooth.Spline()*

A second useful smoother in R is *smooth.spline()*. A general characteristic of smoothers is that complexity is related to the number of points going into the local estimate of the function. A very smooth fit includes many points in each local estimate and is not very complex. The smoothest possible fit to the data is a line, which is also the simplest model. Allowing each individual observation to be its own estimate is the most complex possible fit and would be the least smooth. In the *lowess()* formula, for example, $p = 2(1 + 1/f)$, as f goes up, p goes down.

A spline fit is a function that fits piecewise (in this case, cubic) polynomials to the data. The role of smoothing is less transparent in this case, but there is a way of linking the smoothing to the sample size. For the moment, this is quite ad hoc, after learning about information criteria, a more systematic approach will be available.

For *smooth.spline()*, the complexity of the model is the degrees of freedom. Heuristically, a consistent estimator can be found if it is possible to allow the degrees of freedom to grow with the sample size, but at a slower rate than the sample size. For the moment, $df = 2 + 2 \cdot \sqrt{n}$ will do. This is just done so that the *lowess()* and *smooth.spline()* have the same complexity for comparison purposes. Note that the generic requirement is met; the degrees of freedom grow with sample size, but more slowly than sample size. [As always, it does no harm to explore: help(smooth.spline).]

Returning to the data plotted in Figure 9.2, a *smooth.spline()* fit with about the same complexity as the *lowess()* fit is produced by the following R code:

```
splinefit <- smooth.spline(z$freq, log(z$spec), df = 91.44)
plot(z$freq, log(z$spec), pch = ".")
lines(z$freq, log_pt)
lines(splinefit$x,splinefit$y, lty = 2)
title("Spectrum and smooth.spline() smoother[—]").
```

Spectrum and *lowess()* smoother[---]

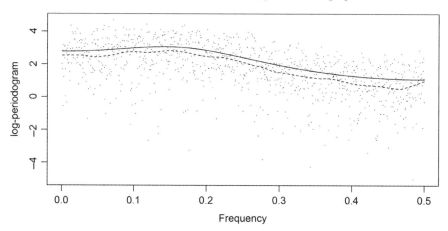

FIGURE 9.2   The *lowess()* fit to the spectrum of an AR(2) model when $f = 1/\sqrt{n}$.

[Explore: It would be insightful to employ the command names(smoothspline) here].

Lowess() and *smooth.spline()* have many similarities and differences (comparing Figures 9.2 and 9.3), and either might perform better in a specific application. In this case, both consistently underestimate the spectrum, but *smooth,spline()* is much "rougher" than *lowess()*.

9.3.3   Kernel Smoothers in *Spec.Pgram()*

The writers of the R function *spec.pgram()* were well aware that the periodogram needed to be smoothed. For completeness, we will discuss the use of the kernel

Spectrum and *smooth.spline()* smoother[---]

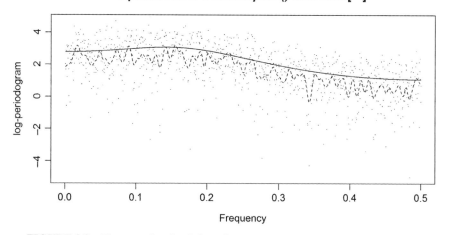

FIGURE 9.3   The *smooth.spline()* fit to the spectrum of the previous AR(2) model.

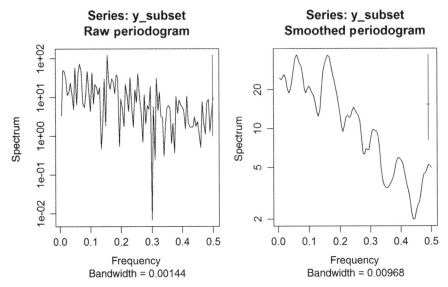

FIGURE 9.4   A periodogram for the first 200 observations from the AR(2) data without smoothing and with smoothing with the *spans()* function.

smoothers used in *spec.pgram()*. Consider the R code zsp <- spec.pgram(y, spans = c(7,3), taper = 0.0, fast = FALSE). This will also smooth a periodogram, see Figure 9.4.

The *spans()* specification uses local weighting to combine nearby values of the periodogram, centered at a frequency, to estimate the spectrum at that frequency. In this case, the "7" indicates an initial weight on seven local values of periodogram (log scale). The middle values all get the same weight and the end values get half the weight of the middle values. The weights sum to 1.0: 0.08333, 0.16667, 0.16667, 0.16667, 0.16667, 0.16667, 0.08333.

In the same manner "3" would have the weights: 0.25, 0.50, 0.25.

The final results, found using zsp$kernel, are: mDaniell(3,1)

coef[-4] = 0.02083, coef[-3] = 0.08333, coef[-2] = 0.14583
coef[-1] = 0.16667, coef[0] = 0.16667, coef[1] = 0.16667
coef[2] = 0.14583, coef[3] = 0.08333, coef[4] = 0.02083

Adding more levels to the span creates a more mound-shaped weighting scheme. For example,

zsp <- spec.pgram(w, spans = c(3,3,3,3), taper = 0.0, fast = FALSE)
zsp$kernel produces the weights:

coef[-4] = 0.003906, coef[-3] = 0.031250, coef[-2] = 0.109375
coef[-1] = 0.218750, coef[0] = 0.273438, coef[1] = 0.218750
coef[2] = 0.109375, coef[3] = 0.031250, coef[4] = 0.003906

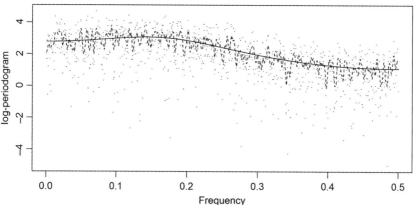

Spectrum and *spans()* smoother[---]

FIGURE 9.5 A smoothing of the periodogram for the AR(2) model using the kernel smoother in *spec.pgram()*.

Given the goal of constructing consistent estimators of the spectrum from the periodogram, there is nothing special about these smoothers. So how can a consistent estimator be constructed in this manner?

The function *spans()* requires odd integers for the procedure to work. A reasonable choice might be to let $r = 2 \cdot floor[\sqrt{n}/2] + 1$. Recall the *floor()* function rounds any value down to the nearest integer. This produces r that is an odd integer close to $\sqrt{n}$. Something like spans = c(r,3,3,3) or spans = c(r,7,5,3) should work reasonably well. Returning to the data from Figures 9.2 and 9.3, the following code produces Figure 9.5 [continuing with z for the periodogram for the AR(2) series y].

```
plot(z$freq,log(z$spec), pch = ".")
lines(z$freq, log_pt)
r <- 2*floor(sqrt(n)/2) + 1
zsp <- spec.pgram(y, taper = 0.0, spans = c(r, 3,3,3), fast = FALSE, plot = FALSE)
lines(z$freq, log(zsp$spec), lty= 2)
title("Spectrum and spans() smoother[—]")
```

This smoothed periodogram (Figure 9.5) looks quite good. The smoothed curve is very close to the true underling curve.

9.4   SMOOTHING THE PERIODOGRAM FOR A SERIES WITH A KNOWN AND UNKNOWN PERIOD

9.4.1   Period Known

Consider the NYC temperatures data (Figure 3.1), which has a period of 12 (annual monthly data). Each smoothing is considered (Figure 9.6). It is useful to know how well a periodogram performs when the period is known.

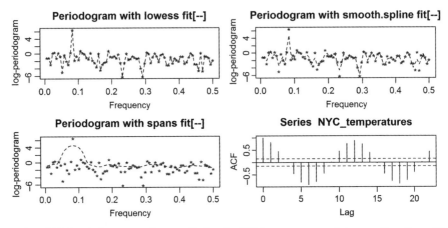

FIGURE 9.6   Various smoothers applied to the NYC temperatures data as well as the *acf()* plot.

In this case, the underlying period is obvious. The data is monthly with a period of 1 year, so $1/12 = 0.0833\ldots$ is the dominant frequency. The smoothers with a factor of $f = 1/\sqrt{n}$ were too smooth, so a factor of $f = 0.5/\sqrt{n}$ was used (further comments on this appear at the end of the chapter).

9.4.2   Period Unknown

Figure 9.7 displays the smoother periodograms for the sunspot data using the same approaches as above to choose the degree of smoothing. A careful examination of the values (Figure 9.7) suggests that both *smooth.spline()* and *spans()* yield a

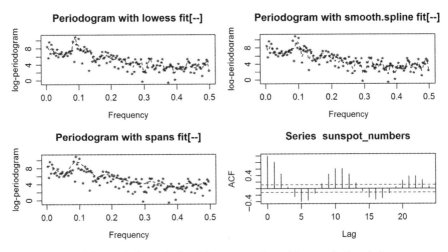

FIGURE 9.7   Periodic analysis of the sunspot data with smoothed periodograms.

maximum for the periodogram at the frequency 0.0968858 suggesting a period of 10.32 (1/0.0968858), while *lowess()* suggests a period of 11.12 years.

9.5 SUMMARY

It is a simple process to produce a smoothed version of the periodogram that is a consistent estimator of the power spectrum. The general rule for producing a smoothed estimate of any continuous, well-behaved function is to smooth the data in such a way that as sample size increases, more observations are included in the estimation of each y value (variance reduction) and the points are chosen from smaller and smaller windows around each x value (bias reduction).

Although it is easy to formulate a general rule that works for infinite sample size, trial and error have been used to produce useful results for finite sample size. In particular, while a window about $\sqrt{n}$ with about $\sqrt{n}$ observations in each window worked reasonably well for the larger sample sizes in the simulated data at the beginning of this chapter, the real data sets produced periodograms that looked good when there were about $2\sqrt{n}$ windows with about $\sqrt{n}/2$ observations in each window (less bias, more variance).

After discussing information criteria, where the tradeoff between bias and variance will be discussed in more detail, some assigned exercises will hint at how more optimal fittings of smoothers to periodograms are possible.

EXERCISES

1. Use the formula for $\hat{C}_k$ to show that for k close to n, only a few lags are used to estimate the value.

2. A failed attempt at a consistent estimator: Box et al. (2008) suggest only $n/4$ correlations should be estimated. Modify the periodogram code to use only floor($n/4$) instead of $n-1$ terms, thus using only "reliable" correlations: $\hat{p}(f) = \hat{C}_0 + 2\sum_{k=1}^{\text{floor}(n/4)} \hat{C}_k \cos(2\pi k f)$. Use the sunspot data. Notice any issues?

3. Invert the formula $df = 2(1 + 1/f)$ to get a formula for $f = g(df)$.

4. (i) Develop the explicit formula for df in *smooth.spline()* and r in *spans()* so that the smoothing of these smoothers is about the same as *lowess()* with $f = \alpha/\sqrt{n}$, where $0.5 < \alpha < 2$.

    (ii) Smooth the NYC temperature data with all three smoothers (spans, smooth.spline, lowess) and three levels of smoothing $f = 0.25/\sqrt{n}, 0.5/\sqrt{n}, 1/\sqrt{n}$ for *lowess()* and the equivalent level smoothing for the other smoothers.

    (iii) Plot each smoothed periodogram (there are nine). Which seems best, which seems worst?

5. Repeat Exercise 4 with the sunspot data.

6. Suppose *lowess()* were used with $f = 1/\sqrt[3]{n}$, would the method produce consistent estimators? What about $f = 1/n^2$? Justify your answer in terms of bias and variance.

7. Assessing the smoothers for MA(2) with $n = 500$, $b_1 = -0.8$, $b_2 = 0.7$:

   (i) Simulate the data.

   (ii) Fit the periodogram to the data and plot it with "log(spec)" as the y variable and "freq" as the x variable.

   (iii) Add a line with the smoothed data (each type of smoother).
   When done, there will be three plots: each will have the same periodogram and true model and a different smoother.

   (iv) Which smoother is best? Which is worst? Explain.

   (v) Turn in your R code with each plot.

8. Repeat the steps in Exercise 7 for MA(2) with $n = 200$, $b_1 = 0.3$, $b_2 = -0.3$.

9. Find the weights associated with each of these smoothing approaches (you do not need to find these by hand):

   (i) sp3 <- spec.pgram(y,taper = 0, fast = FALSE, spans=c(3))

   (ii) sp33 <- spec.pgram(y,taper = 0, fast = FALSE, spans=c(3,3))

   (iii) sp5 <- spec.pgram(y,taper = 0, fast = FALSE, spans=c(5))

   (iv) sp35 <- spec.pgram(y,taper = 0, fast = FALSE, spans=c(3,5))

10. Melbourne temperatures ("Melbmax.txt" from the Exercises folder):
    Develop smoothed periodograms for the Melbourne daily temperature data. What is the correct dominant frequency? Which smoothing method (spans, smooth.spline, lowess) produces the best estimate of the correct dominant frequency? Comment on the graphs.

11. lynx data ("LYNX.txt" from the Exercise folder):
    Using the smoothing methods, estimate the period, in years, of the lynx data. Comment on each method.

10

A REGRESSION MODEL FOR PERIODIC DATA

10.1 THE MODEL

A general periodic model is of the form, $y(j) = \mu + M \cdot \cos(2\pi[j/k + \phi])$, where j is integers indexing time. The mean, μ, shifts the function up and down, M changes the amplitude, and ϕ shifts (sometimes called the phase) the location of the maximum and minimum. The function repeats itself every k periods. For annual data, k would be 12 for monthly data and 52 for weekly data. Figure 10.1 displays the function $y_1(j) = 2 + 8 \cdot \cos(2\pi[j/12 + .25])$ in solid lines and $y_2(j) = 4 + 4 \cdot \cos(2\pi[j/24 + .70])$ in dashed lines.

The most basic periodic models are of the form $y(j) = \mu + M \cdot \cos(2\pi[j/k + \phi]) + \varepsilon_j$, where the errors are white noise. However, the time series may be more complex requiring a linear combination of periodic functions and the error may not be white noise.

For this most basic model, it is assumed that k is known, and μ, M, and ϕ must be estimated. The problem $\min \sum [y(j) - \mu - M \cdot \cos(2\pi[j/k + \phi])]^2$, where μ, M, and ϕ are unknown, is not an OLS (ordinary least squares) problem. Why (hint: take the partial derivatives and set them equal to zero)?

However, there is a trick that allows this to be solved as an OLS problem (Bloomfield, 2000). Let $y(j) = \mu + B \cdot \cos(2\pi j/k) + C \cdot \sin(2\pi j/k)$ where $B = M \cdot \cos(2\pi\phi)$ and $C = -M \cdot \sin(2\pi\phi)$. The problem can be restated as the OLS problem:

$$\min \sum [y(i) - \mu - B \cdot \cos(2\pi j/k) - C \cdot \sin(2\pi j/k)]^2.$$

Basic Data Analysis for Time Series with R, First Edition. DeWayne R. Derryberry.
© 2014 John Wiley & Sons, Inc. Published 2014 by John Wiley & Sons, Inc.

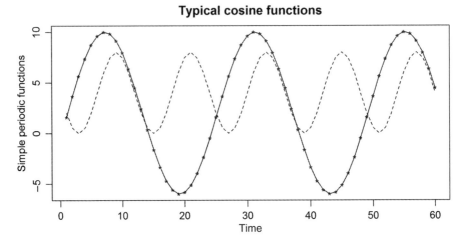

FIGURE 10.1   Cosine functions can vary in amplitude, mean, period, and phase.

Fitting the model to the data using OLS, the matrix representation of the system is

$$
\begin{pmatrix} y(1) \\ y(2) \\ \dots \\ y(n) \end{pmatrix} = \begin{bmatrix} 1 & \cos(2\pi/k) & \sin(2\pi/k) \\ 1 & \cos(2\pi2/k) & \sin(2\pi2/k) \\ \dots & \dots & \dots \\ 1 & \cos(2\pi n/k) & \sin(2\pi n/k) \end{bmatrix} \begin{pmatrix} \mu \\ B \\ C \end{pmatrix}
$$

After the problem has been solved $\hat{M}$ and $\hat{\phi}$ are found as follows:

$$
\hat{B}^2 + \hat{C}^2 = \hat{M}^2\cos^2(2\pi\hat{\phi}) + M^2\sin^2(2\pi\hat{\phi}) = \hat{M}^2 \quad \text{so } \hat{M} = \sqrt{\hat{B}^2 + \hat{C}^2}
$$

and (yikes), producing an estimate of ϕ is tricky (Bloomfield, 2000, p. 13).

$$
2\pi\hat{\phi} = \begin{cases} \tan^{-1}(-\hat{C}/\hat{B}) & \hat{B} > 0 \\ \tan^{-1}(-\hat{C}/\hat{B}) - \pi & \hat{B} < 0, \hat{C} > 0 \\ \tan^{-1}(-\hat{C}/\hat{B}) + \pi & \hat{B} < 0, \hat{C} < 0 \\ -\pi/2 & \hat{B} = 0, \hat{C} > 0 \\ \pi/2 & \hat{B} = 0, \hat{C} < 0 \\ ?? & \hat{B} = 0, \hat{C} = 0 \end{cases}
$$

10.2   AN EXAMPLE: THE NYC TEMPERATURE DATA

10.2.1   Fitting a Periodic Function

Recall that the NYC temperature data is monthly, so it has period $k = 12$, as can be easily seen in Figure 3.1.

The steps to fit the simple periodic model such as the NYC temperature data:

(i) form a column for the cosine values and a column for the sine values;
(ii) form a column of "ones" to model the intercept;
(iii) combine these columns to form a matrix representing the least squares problem;
(iv) form the OLS problem in R and solve using *lm( )*, this will produce

$$\hat{\mu}, \ \hat{B} = \hat{M}\cos(2\pi\hat{\phi}) \text{, and } \hat{C} = -\hat{M}\sin(2\pi\hat{\phi});$$

(v) use $\hat{B}$ and $\hat{C}$ to find $\hat{M}$ and $\hat{\phi}$;
(vi) use the model to answer questions of interest: In this case, a simple summary of the data is presented. What is the normal range of average temperatures? Which month is the hottest?

By now, this code should be clear, after scanning the data, labeled y_NYC:

```
time <- c(1:n)
col_1 <- rep(1,n)
col_c <- cos(2*pi*time/12)
col_s <- sin(2*pi*time/12)
X <- cbind(col_1,col_c,col_s)
fit_NYC <- lm(y_NYC ~ -1+ X)  # since the intercept terms are in X,
#not intercept is fitted
```

The output is in Table 10.1.

From Table 10.1, the initial fit is $T = 15.3 - 2.93 \cdot \cos(2\pi j/12) - 2.54 \cdot \sin(2\pi j/12)$, for $j = 1, 2, 3 \ldots$ and where T is mean monthly temperature.

TABLE 10.1   The Initial Fit of a Periodic Function to the NYC Temperature Data

lm(formula = y_NYC ~ -1 + X)
Coefficients:

	Estimate	SE	*t*-value	Pr(>|*t*|)
Xcol_1	15.33289	0.05203	294.70	<2e−16
Xcol_c	−2.92704	0.07358	−39.78	<2e−16
Xcol_s	−2.53766	0.07358	−34.49	<2e−16

Residual SE, 0.6744 on 165 df; multiple R^2, 0.9982; adjusted R^2, 0.9981; F-statistic, 2.987e+04 on 3 and 165 df; p-value, < 2.2e−16

New York city monthly temperatures

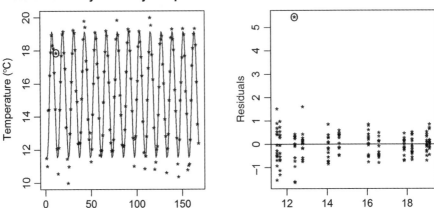

FIGURE 10.2   The New York City temperature data has a clear outlier.

10.2.2   An Outlier

The model does seem to fit the data well except for one not-so-obvious outlier: plot(time,y_NYC), lines(time,fit_NYC$fitted.values).

The *lm( )* function in R produces residuals, and here the outlier stands out clearly (Figure 10.2, right panel). After looking at the residual plot, it is easy to return to the left panel of Figure 10.2 and spot the outlier, a very large value early in the graph where the model predicts a low value.

The outlier is at the 12th observation (December 1946). The data suggests that the average temperature for December 1946 was y_NYC[12] 17.862°C. This seems overwhelmingly implausible. Is this possible? A quick internet search showed two things: (i) Nothing was found suggesting anything unusual about the temperature in NYC in December 1946—this would have been news and (ii) the NOAA website [(http://www.ncdc.noaa.gov/temp-and precip/timeseries/index.php?parameter= tmp&month=12&year=1946&filter=1&state=30&div=0) accessed January 24, 2013] gives the monthly average as −2.11°C (converted from 28.2°F) for all of New York state.

10.2.3   Refitting the Model with the Outlier Corrected

A file "NYC temp adj.txt" in the Examples folder has the value for December 1946 replaced with the average temperature for all of the other December values in the data set. This creates a value of 12.08746°C for December 1946. When using this data set, the sample size is effectively 167 ($n − 1$), since this is an interpolated, not an observed, value. This model will be used from here forward when analyzing the NYC temperature data. The refitted model is displayed in Table 10.2.

TABLE 10.2   The Corrected Fit for the NYC Temperatures

Coefficients:

	Estimate	SE	t-value	Pr(>|t|)
Xcol_1	15.29851	0.04019	380.66	<2e–16
Xcol_c	–2.99579	0.05684	–52.71	<2e–16
Xcol_s	–2.53766	0.05684	–44.65	<2e–16

Based on Table 10.2, the fitted model is

$$T = 15.3 - 3.0 \cdot \cos(2\pi j/12) - 2.54 \cdot \sin(2\pi j/12) \text{ , for } j = 1, 2, 3, \ldots$$

Straightforward inversion from the formulas in the last section produces

$$\hat{M} = \sqrt{B^2 + \hat{C}^2} = 3.93 \text{ and } 2\pi\hat{\phi} = \tan^{-1}(-\hat{C}/\hat{B}) + \pi \text{ , so } \hat{\phi} = 0.388.$$

Since the data is monthly, beginning in January, $j = 1, 13, 25\ldots$ are January indices. The cosine function, $\cos(2\pi[j/k + \phi])$, is maximized at 0 and 2π. So, to find the month that maximizes temperature, we can solve for $j/12 + \phi = 1$, which produces $j = 7.34$. Unsurprisingly, the hottest period is between July (7) and August (8).

Summarizing the data analysis: The average temperature in NYC is about 15.3°C, with a seasonal variation of $\pm$ 3.9°C. The hottest period is early to mid-July and the coldest times (based on the symmetry of the model) are mid- to late January.

Note: The conclusions are based on the model and the data. The model, not the data, forces the maximum and minimum to be 6 months apart. If this is not true of the data, the model will fit poorly for some parts of the year (more complex models are coming).

The $acf()$ plot in Figure 10.3 suggests that the residuals may not be much different from white noise (more concrete methods of assessing this claim are coming). If that is the case, predictions, prediction intervals, and confidence intervals are routine.

Examination of the residuals (with and without the outlier, in this case) will become the basis diagnosing AR(m) and/or MA(l) structure in the noise. For time series data, white noise residuals will be the exception, not the rule.

10.3   COMPLICATIONS 1: CO$_2$ DATA

A simple periodic model rarely fits the data well, although a variety of extensions of this model get the job done. In the next few sections, several typical complications are described. The complications suggest fitting one of the several more complex models. This in turn suggests model selection—making decisions about which model to use for the data. After a discussion of model selection (Chapter 11), it will be possible to continue with more complex models (Chapter 12).

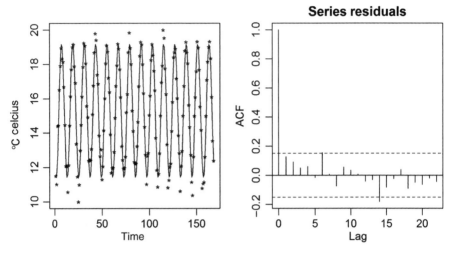

FIGURE 10.3 The fitted model with the outlier corrected.

Recall the Mauna Loa CO_2 data displayed in Figure 3.4. Obviously, the data displays both a trend and a periodic component: Signal = trend + periodic pattern. However, the trend could be linear, quadratic, or more complex and even the periodic component could be more complex than the simple one presented in this chapter.

10.4 COMPLICATIONS 2: SUNSPOT NUMBERS

Recall the sunspot data presented in Figures 3.2 and 10.4 (left panel). Although the data is periodic, there are a number of other issues: A sine or cosine function is symmetric above and below the mean. This data seems to have rounded minima and sharp-peaked maxima. A trigonometric approach will better fit the data if this problem can be (at least partially) remedied. A histogram as a tool for identifying transformations of the data would seem to be of limited value in this context, as the mean of the data is oscillating wildly, nevertheless, such histograms are often somewhat informative.

Either a logarithmic or a square-root transformation of the data would produce a new series more amenable to fit a simple trigonometric model. It is often the case that periodic time series have rounded minima and sharp-peaked maxima. In these cases, the square root or logarithmic transformation seems to work well most of the time.

Based on Figure 10.5, it appears that square-root transformation (right panels) corrects for the right skew in the data well, but the logarithmic transformation over-corrects (left panels). To the trained observer, this is obvious from the plots over time, and the histograms are a far less-effective way of capturing this information (and

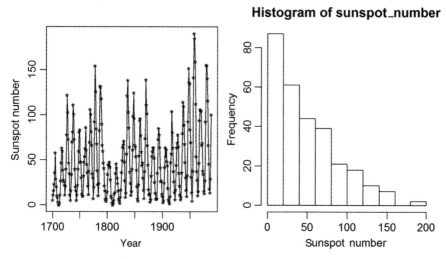

FIGURE 10.4   Right skew and unknown period.

could even be misleading in some cases), but those familiar with histograms and new to periodic data will find the histograms useful (until their "eyesight" improves).

A second complication is that the true period is unknown. Although, in the previous chapter, it was possible to produce a crude estimate of the period, it is desirable to optimize this estimate either by optimal smoothing of the periodogram or by solving a nonlinear problem where the period is free to vary over time.

Finally, the period and/or amplitude may vary. If this is the case, developing a sound method of characterizing this variation would be illuminating. It will be possible to fit a model to the data even if the period, phase, amplitude, and mean vary.

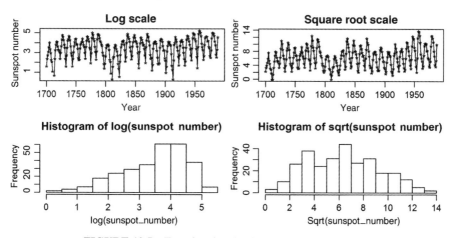

FIGURE 10.5   Transforming the data to address right skew.

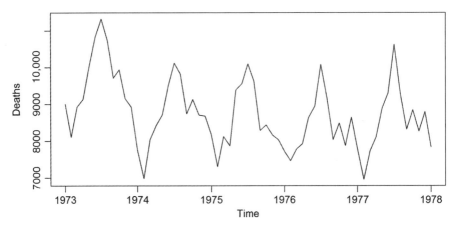

FIGURE 10.6   Accidental deaths in the US from January 1973 to December 1978.

10.5   COMPLICATIONS 3: ACCIDENTAL DEATHS

The number of accidental deaths ("Accidental deaths.txt" from the Examples folder or DataMarket–Time Series Data Library–Demography) displays a count of accidental deaths in the United States (Figure 10.6).

In this case, there is no trend, and the period is clearly 12 (monthly data), nevertheless a simple periodic function does not capture all of the subtlety in the shape. There is an approach for adding complexity to the simple periodic model, based on Fourier series, but a "best" model must be chosen from a range of models varying in both complexity and fit.

10.6   SUMMARY

There are many possible complications in fitting a model to data, although not all complications have been covered, model selection is the key to addressing almost all complications like those so far seen. Recall the overall goal. To fit a signal to the data, then fit a model [AR(m) and/or MA(l) models] to the noise so that statistical inference is possible. Until it is possible to fit reasonable signals to the data, it is not even possible to produce residuals and assess the nature of the noise.

EXERCISES

1. **(i)** Show that the partial derivatives of min $\sum [y(j) - \mu - M \cdot \cos(2\pi[j/k + \phi])]^2$ do not form a system of linear equations.

   (ii) Show that the partial derivatives of min $\sum [y(i) - \mu - B \cdot \cos(2\pi j/k) - C \cdot \sin(2\pi j/k)]^2$ do form a system of linear equations.

2. In the following four cases, determine M and ϕ from B and C:

   (i) $B = 3.7, C = 2.1$

  (ii) $B = -2.1, C = -7.1$

 (iii) $B = -3.2, C = 1.3$

 (iv) $B = 2.2, C = -2.3$

3. Let $M \cdot \cos(2\pi [ft + \phi]) = B \cdot \cos(2\pi ft) + C \cdot \sin(2\pi ft)$.
   Show that $B = M \cdot \cos(2\pi\phi)$ and $C = -M \cdot \sin(2\pi\phi)$.

4. The impact of one outlier in 168 observations: fit the simple periodic model to the New York temperature data with the outlier still included: (i) recompute values of the summary paragraph; (ii) compare the R^2 values; (iii) create the *acf()* plot of the residuals; (iv) based on this analysis, what is the impact of a single outlier in a data set of 168 observations?

5. Consider the Melbourne maximum daily temperature data ("Melbmax.txt" from the Exercises folder). Fit a simple periodic model to this data and write up a conclusion similar to the summarizing paragraph given for the NYC temperature data (average rainfall, range of rainfall, when is the hottest and coldest time).

   How does the weather in Melbourne compare to the weather in NYC?

6. For the lynx data ("LYNX.txt" from the Exercises folder):

  (i) Plot the original data versus time. Does it appear that the minima are curved/rounded and the maxima are sharply peaked?

 (ii) Plot the data (versus time) with both the logarithmic transformation and the square-root transformation and examine whether either (or both) transformation makes the data more amenable to a trigonometric analysis.

7. Monthly milk production in pounds per cow from January 1962 to December 1975 ("Milk production.txt" from the Exercises folder). Plot this data versus time. What complications do you foresee here (what type of signal would fit this data)?

8. Plot the number of deaths and serious injuries in UK road accidents each month from January 1969–December 1984 versus time—a seatbelt law was introduced in February 1983 ("UK road deaths.txt" in the Exercises folder or DataMarket–Time Series Data Library–Transport and Tourism). There are two columns of data with a header, use *read.table()* to retrieve the data. What complications do you foresee in modeling this data?

11

MODEL SELECTION AND CROSS-VALIDATION

11.1 BACKGROUND

A number of situations will be presented in which several possible models are proposed to fit a set of data. In these situations, both hypothesis testing and information criteria can be used for model selection. Both hypothesis testing and information criteria are rooted in the same basic idea: a more complex model will generally fit the data better than a simple model, so that the complexity of models and how well they fit the data must be balanced in making a decision about which model best fits the data. Two other strands enter into the discussion: models may not be nested, and picking the best model requires clarification. By best model, does one mean the model most likely to have generated the data, or the model most likely to make good future predictions?

For linear models with normal errors, fit can be quantified as SSE $= \sum (y_j - \hat{y}_j)^2$, which is just the sum of the squared residuals. For all models, complexity can be characterized by the dimension of the parameter space p. Although this sounds quite intimidating, for frugally parameterized models (always the case in this book), it is just the number of parameters estimated using OLS. Hence, every model selection problem contains a table with the values of SSE and p for each model. Different approaches to model selection just use these values in different ways. Once the principles of model selection are understood, extending these ideas to time series data will be trivial.

Basic Data Analysis for Time Series with R, First Edition. DeWayne R. Derryberry.
© 2014 John Wiley & Sons, Inc. Published 2014 by John Wiley & Sons, Inc.

11.2 HYPOTHESIS TESTS IN SIMPLE REGRESSION

Consider the model $y_j = \beta_0 + \beta_1 x_j + \varepsilon_j$, with independent normal errors.

The most common test in this situation is $Ho : \beta_1 = 0$ versus $Ha : \beta_1 \neq 0$. In other words, is there evidence of a linear trend in the plot of x versus y? A t-test of the form $t = (\hat{\beta}_1 - 0)/s_{\hat{\beta}_1}$ with $n - 2$ degrees of freedom is appropriate. It is common when discussing such tests in the usual regression framework to generalize these results to the extras sums of squares F-test. However, in the case of model selection and information criteria it makes more sense to generalize these results to likelihood ratio tests.

Following the usual approach, simulation of data with known properties will usually be used to construct examples when demonstrating standard theory. For example, consider a case where a line produced the data and both the usual t-test and the likelihood ratio test are used to verify this.

```
# simulate 50 observations for a simple regression with normal errors.
# and standard deviation 2.
# Because several simulations will be performed in this chapter.
# extensions on the values _1, _2,_3, etc., will be used to keep them distinct.
sigma_1 <- 8.0.
n_1 <- 50.
err_1 <- rnorm(n_1,0,sigma_1).
x_1 <- c(1:n_1).
y_1 <- 3.1 - 0.5*x_1 + err_1 # the true model is of the form y = 3.1 − 0.5x + ε.
# fit the data using lm().
fit_1 <- lm(y_1 ~ x_1).
```

The fitted model is displayed in Table 11.1.

The slope (β_1) is estimated to be -0.62227 $(\hat{\beta}_1)$. The t-test statistic is $\frac{-0.62227-0}{0.08641} = -7.201$ with 48 $(50 - 2)$ degrees of freedom. There is overwhelming evidence $(p$-value $\approx 3.62 \times 10^{-9})$ a line fits the data compared to a simple mean (the null hypothesis).

A second approach, one that generalizes to more complex situations, is the likelihood ratio test. Nested models are models in which one the simpler is a special

TABLE 11.1   The (Partial) Result of Fitting the Model in R, Parameter Estimates

Call: lm(formula = y_1 ~ x_1)
Coefficients:

	Estimate	SE	t-value	Pr(>|t|)
(Intercept)	5.90408	2.53197	2.332	0.024
x_1	-0.62227	0.08641	-7.201	3.62e−09

Residual SE: 8.818 on 48 df; multiple R^2: 0.5193; adjusted R^2: 0.5093; F-statistic: 51.85 on 1 and 48 df; p-value: 3.619e–09.

case of the more general model. For examples, in the case of regression the simpler model (reduced model) is a simple mean and the more complex model (full model) is a line. The reduced model can be viewed as a special case of the full model in that the reduced model is the full model when $\beta_1 = 0$.

In general, a full model is associated with the alternative hypothesis and is a model where all parameters are free to vary in the OLS formulation (assuming normal errors). The reduced model is found by fixing at least one of the free parameters in the full model at some predetermined value (usually, but not always, zero). It is said the models are nested, and the reduced model is always a special case of the full model. Let the full model have sum of squared residuals SSE_F and p_F free parameters, while the reduced model has sum of squared residuals SSE_R and free parameters $p_R (< p_F)$. The p-value for a likelihood ratio test for such models, in the normal setting, is based on the claim that, asymptotically, when the null hypothesis is true: $\chi^2_{p_F - p_R} \sim n \cdot \log_e(SSE_R) - n \cdot \log_e(SSE_F)$. In other words, the p-value for the likelihood ratio test can be found by comparing the computed test statistic to a chi-square distribution whose degrees of freedom are the difference in the number of free parameters in each model.

The computation of SSE for a model with a simple mean is quite common, as this is the usual simplest possible model for comparison purposes. In this case sum of the squared residuals have a special designation $SST = \sum(y_j - \bar{y})^2 = s_y^2(n - 1)$.

var(y_1)*(50-1) # SST	7764.261.
sum(fit_1$residuals^2) # SSE for the full regression model	3732.278.

Note: $R^2 = 1 - SSE/SST = 0.5193$, which can be verified in the regression output (Table 11.1).

The same numbers are produced in Table 11.2 using the command anova(fit_1). In this case SST = 4032.0 + 3732.3 = 7764.3 and SSE = 3732.3.

Based on the results from Table 11.2, the simplest version of a common pattern begins to form. Table 11.3 is a tabular display of all candidate models with their fit (SSE) and complexity (p). Tables such as this, sometimes with as many as 8–10 candidate models, will appear prominently in the rest of this book.

The test statistic for the likelihood ratio test, based on information in Table 11.3 and the sample size, is $50 \cdot \log_e(7764.3) - 50 \cdot \log_e(3732.3) = 36.63$. In this case, with a chi-square of just one degree of freedom (2–1), the p-value is roughly 1.64×10^{-9}. This approximate p-value has about the same magnitude as the more exact t-test

TABLE 11.2   The ANOVA (Analysis of Variance) Table for the Regression anova(fit_1)

Response: y_1

	df	Sum Sq.	Mean Sq.	F-value	Pr($>F$)
x_1	1	4032.0	4032.0	51.855	3.619e–09
Residuals	48	3732.3	77.8		

TABLE 11.3   Summary of the Relevant Information for the Competing Models

Model	SSE	Complexity (p)
One mean (reduced model)	7764.3	1
Line (full model)	3732.3	2

p-value. Obviously, there is overwhelming evidence that a line is a better fit to the data than a simple mean.

11.3   A MORE GENERAL SETTING FOR LIKELIHOOD RATIO TESTS

Consider a sequence of nested models:

Model 1—a simple mean: $y_j = \beta_0 + \varepsilon_j$.
Model 2—a straight-line fit: $y_j = \beta_0 + \beta_1 x_j + \varepsilon_j$.
Model 3—a quadratic fit: $y_j = \beta_0 + \beta_1 x_j + \beta_2 x_j^2 + \varepsilon_j$.

Since these models are nested, the "best" fit is selected by performing a sequence of tests:

Test no. 1: (null) Model 1 versus (alternative) Model 2.
Test no. 2: (null) Model 2 versus (alternative) Model 3.

Having already used a straight-line model to simulate data (later a quadratic model will be used), it is possible to first demonstrate the idea of sequential tests by producing a quadratic fit to the current data (Table 11.4):

```
x_1_2 <- x_1*x_1.
qfit_1 <- lm(y_1 ~ x_1 + x_1_2).
anova(qfit_1).
```

Table 11.3 is now expanded to include the new information, forming Table 11.5.
Using Table 11.5, it is possible to perform the second test: comparing linear versus quadratic model $50 \cdot \log_e(3732.3) - 50 \cdot \log_e(3632.9) = 1.35$. This results in

TABLE 11.4   A Quadratic Model Fitted to the Same Data

Response: y_1

	df	Sum Sq.	Mean Sq.	F-value	Pr($>F$)
x_1	1	4032.0	4032.0	52.1770	3.75e–09
x_1_2	1	100.3	100.3	1.2986	0.2603
Residuals	47	3631.9	77.3		

TABLE 11.5   The Additional Model Included in the Same Format

Model	SSE	Complexity (p)
One mean	7764.3	1
Line	3732.3	2
Quadratic	3631.9	3

a p-value of 0.2453. In other words, a linear model fits the data much better than a single mean, but a quadratic model does not do much better than a linear model. The combined results of the two hierarchical tests are that a line best fits the data.

In fact, it is obvious that a straight-line fit is best when the data is plotted (Figure 11.1):

```
plot(x_1, y_1)                    lines(x_1, fit_1$fitted).
lines(x_1, qfit_1$fitted, lty = 2).
```

It should be noted that it is only obvious which model best fits the data when the sample size is moderate (say, $n = 30$ to 2000). For really large and small data sets, the results of model selection may not match intuition.

For completeness, consider the situation where the true signal is known to be quadratic:

```
sigma_2 <- 0.5.
n_2 <- 80.
err_2 <- rnorm(n_2,0,sigma_2).
x_2 <- c(1:n_2)/n_2   # this produces x values uniformly on the interval 0 to 1.
```

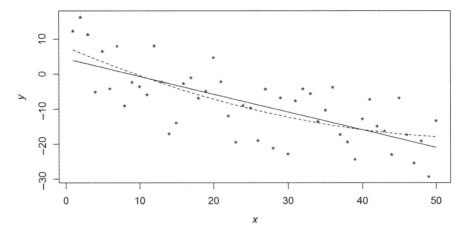

FIGURE 11.1   Models fitted to data known to have a linear underlying model.

TABLE 11.6   An ANOVA Table When the Underlying Model is Quadratic

Response: y_2

	df	Sum Sq.	Mean Sq.	*F*-value	Pr(>*F*)
x_2	1	3361.0	3361.0	16,804.25	<2.2e−16
x_2_2	1	185.0	185.0	924.86	<2.2e−16
Residuals	77	15.4	0.2		

```
y_2 <- 7.1 + 2*x_2 + 20*x_2*x_2 + err_2   # a quadratic model.
# fit the data using lm().
fit_2 <- lm(y_2 ~ x_2)          # fitting a linear trend.
x_2_2 <- x_2*x_2.
qfit_2 <- lm(y_2 ~ x_2 + x_2_2)      # fitting curvature.
anova(qfit_2).
```

This R code results in Table 11.6.

From the ANOVA table, $SST = SSE_{mean} = 3361.0 + 185.0 + 15.4 = 3561.4$, $SSE_{line} = 185.0 + 15.4 = 200.4$, and $SSE_{quadratic} = 15.4$. These are sequential sums of squares. The sum of all the sums of squares in an R ANOVA table is $SST = var(y) \cdot (n - 1)$. (Exercise: What would this ANOVA table look like if only a line were fitted to the data?) These computations allow for a more useful display of the model fits found in Table 11.7. The sample size is always one more than the total degrees of freedom in the ANOVA table ($n = 1 + 1 + 77 + 1 = 80$).

From Table 11.7: the two tests are

Test no. 1: mean versus line $80 \cdot \log_e (3561.4) - 80 \cdot \log_e (200.4) = 230.21$; the *p*-value is very small and a line fits the data much better than a single mean.

Test no. 2: a line versus a quadratic curve $80 \cdot \log_e (200.4) - 80 \cdot \log_e (15.4) = 205.28$; the *p*-value is again quite small and a quadratic function fits the data better than a line.

The results are quite obvious when looking at the data.

All roads lead to Rome: Of course, relevant plots of the data (Figure 11.2) reveal the same conclusion as the hypothesis tests.

TABLE 11.7   The Competing Models with Simulated Quadratic Data

Model	SSE	Complexity (*p*)
One mean	3561.4	1
Line	200.4	2
Quadratic	15.4	3

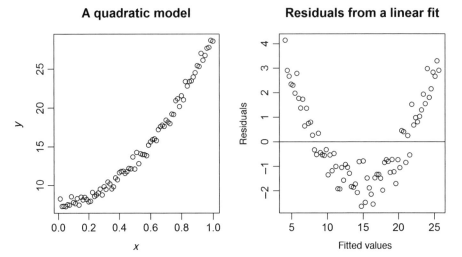

FIGURE 11.2   Both the plot and the residuals show curvature in the data.

It is worth remembering that statistical tests do not always get the right conclusion for simulated data. The smaller the sample size and the smaller the signal relative to the noise, the less often the correct conclusion is reached about the form of the signal. For real data the larger questions are whether any of the models is "right" and whether identification of the right model is even the objective, and if not, what is the objective?

11.4   A SUBTLETY DIFFERENT SITUATION

In calibration problems, there is often an interest in whether two sets of measurements are approximately the same over a wide range of measurements. Perhaps one measurement is much cheaper to make and there is an interest in whether the new, less expensive measurement can be used in place of the older, standard, more expensive procedure.

There are many cases in which we wish to assess whether a 45-degree line (intercept zero, slope one) is the best fit to the data. At first, it might seem the simple test.

$Ho : \beta_0 = 0$, $\beta_1 = 1$ versus $Ho : \beta_0 \neq 0$, $\beta_1 \neq 1$ is all that is required. However, a large p-value does not guarantee that a 45-degree line is the best fit to the data. If neither model fits the data well, a large p-value could also occur.

Consider a simulation that produces the output y_3, which is just x_3 with added white noise. Here a 45-degree line is generating the data (the results are in plotted in Figure 11.3).

A calibration problem

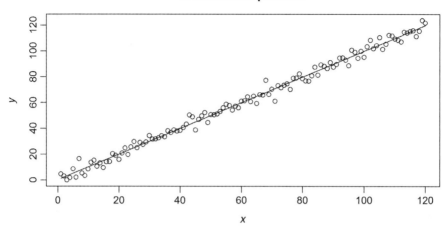

FIGURE 11.3   Simulated data produced by a 45-degree line (slope = 1, intercept = 0) with added noise.

```
sigma_3 <- 3.0.
n_3 <- 120.
err_3 <- rnorm(n_3,0,sigma_3).
x_3 <- c(1:n_3).
y_3 <- x_3 + err_3.
# fit the data using lm().
fit_3 <- lm(y_3 ~ x_3).
```

There are really two questions involved: Does a straight line fit the data? And if a line fits the data, is a 45-degree line the best line? Whenever the 45-degree line is one of the models the nested modeling approach breaks down.

Model 1—a simple mean: $y_j = \beta_0 + \varepsilon_j$.
Model 2—45-degree line: $y_j = x_j + \varepsilon_j$.
Model 3—a straight-line fit: $y_j = \beta_0 + \beta_1 x_j + \varepsilon_j$.

Models 1 and 2 are both special cases of model 3, but there is no nesting relationship between models 1 and 2. In this case, a likelihood ratio testing approach may require some clever, ad hoc modifications. Of course the proper conclusion is obvious from a tabular display of the information (Table 11.8).

For the 45-degree line, SSE = $\sum (x_j - y_j)^2$ (why?). It seems the 45-degree line does best fit the data, it is both simplest and has an SSE (Table 11.8) value almost as small as the general line. In fact, if the 45-degree line had the smallest SSE, it would both be simplest and best fitting and would be chosen without any formal test.

TABLE 11.8   The Model Selection Information for Non-Nested Models

Model	SSE	Complexity (p)
One mean	14,824.7	1
45-degree line	1215.3	0
General line	1193.0	2

However, without a nested relationship, the way to proceeded is unclear. Of course, it is obvious that the 45-degree line best fits the data, but perhaps there is an interest in how much better it fits the data, than the other models.

11.5   INFORMATION CRITERIA

There is an approach to model selection that does not require that models be nested. This approach considers only the complexity of the model and the size of the residual sums of squares. Although the theoretical derivations are quite sophisticated, these approaches can be understood in simple terms—find an optimal way to balance complexity and fit. In the special case of normal error models, the criteria are defined as

Akaike's information: $\mathrm{AIC} = n \cdot \log_e(\mathrm{SSE}) + 2(p + 1) + C,$

Schwarz information: $\mathrm{BIC} = n \cdot \log_e(\mathrm{SSE}) + p \cdot \log_e(n) + C,$

where C is an arbitrary constant—the numerical values are not intrinsically meaningful, but the differences are, so the fact that different authors have added various constants, which are ignored here, is not critical. Notice that the information approach is the negative of (two times) the log-likelihood with a penalty for complexity added.

In these equations, p is the complexity of the model, the number of parameters estimated. The AIC value also adjusts for the number of nuisance parameters (Burnham and Anderson, 2002, p. 12), which in this case is just the estimated variance.

A technical point is that all parameter estimates should be maximum likelihood estimates. When other estimates are used, the user should be aware that the AIC and BIC values are approximate.

For small sample sizes *and normal data*, there is a slightly better version of AIC, which corrects for (small) sample size: $\mathrm{AIC}_c = n \cdot \log_e(\mathrm{SSE}) + 2n(p + 1)/(n - p - 2) + C.$

This version has a nice feature in that it automatically penalizes models for fitting too many parameters relative to the sample size. This approach will occasionally be used in this book when sample sizes are small.

In each case, the goal is to minimize the criterion. It is known that AIC is asymptotically efficient, that is, it maximizes predictive accuracy (Anderson and Burnham, 2002; Aho et al., 2013). *If this approach is used to select among models, as sample*

TABLE 11.9   The Previous Models with Information Criteria ($n = 120$)

Model	SSE	Complexity (p)	AIC	BIC
One mean	148,247.7	1	1431.98	1433.58
45-degree line	1215.3	0	854.43	852.37
General line	1193.0	2	856.11	859.68

size increases, it will select the model that is most useful for making accurate future predictions, even when no model is "right."

It is known BIC is (asymptotically) consistent (Kass and Raftery, 1995; Aho et al., 2013). If this approach is used, as sample size increases, it will select the model that generated the data if that model is one of the candidate models.

No method can be both efficient and consistent (Yang, 2005) for the kinds of regression models routinely employed in this book, they are different criteria, and a choice must be made about the goals of the model selection process. AIC is already widely used in time series, seems more appropriate for real data (as opposed to simulated data). For simulated data when one of the proposed models is usually exactly correct (as in the examples above), BIC makes sense. On the other hand, for real data there is little reason to believe, in the opinion of this author, that any of the proposed models actually generated the data: "All models are wrong, but some are useful" (widely attributed to George Box).

In this case the 45-degree line, by virtue of the lowest AIC and BIC value (Table 11.9), is the model chosen. In this case, because the models included the correct model, BIC should be used for model selection.

How big a difference in AIC or BIC, between two models, is big? Consider the quantity $\exp[-0.5 \cdot \Delta IC]$, where ΔIC is a difference in the information criteria value between any two models for either information criteria. This quantity can be viewed as the relative likelihood of one model over another, given the data (Anderson and Burnham, 2002, Sections 2.6–2.10; *The Statistical Sleuth*, 2002, Sections 12.4–12.5). The reasoning involved has a Bayesian feel and may not be to everyone's taste but does provide some guidance as to the magnitude of the associated differences.

These same computations can also be viewed as likelihood ratios that have been adjusted for differences in the number of parameters between the models. It is well known that likelihood ratios are an important kind of evidence (Edwards, 1992; Royall, 1997). In this case, the trust one has in the final ratio depends on the trust one has in the adjustment for unequal numbers of parameters.

With regard to the one mean model versus the model of a general line, AIC suggests, $\exp([1432.98 - 856.11]/2) = 1.844 \cdot 10^{125}$ and BIC gives a similar result. This indicates that the general line is astronomically more likely that the one mean model.

With regard to the 45-degree line versus the general line, AIC suggests $\exp([856.11 - 854.43]/2) = 2.32$ and BIC suggests $\exp([859.68 - 852.35]/2) = 39.1$. In either case, the 45-degree line is much more plausible than the general line, based on the computed values.

TABLE 11.10   Information Criteria for the First Data Set (Table 11.5, $n = 50$)

Model	SSE	Complexity	AIC	BIC
One mean	7764.3	1	451.86	451.78
Line	3732.3	2	417.24	419.06
Quadratic	3631.9	3	417.88	421.61

Revising the first example, it is possible to compare models using information criteria.

In this case, both methods pick the straight line by a wide margin over a single mean model, and by a much smaller margin over the quadratic model (Table 11.10).

To be a bit irreverent, information criteria is great stuff even if it is not quite clear what it is. Nevertheless, information criteria are useful, they are here to stay, and over time, sophisticated statisticians will develop a "feel" for what they are and are not.

11.6   CROSS-VALIDATION (DATA SPLITTING): NYC TEMPERATURES

11.6.1   Explained Variation, R^2

The explained variation, $R^2 = 1 - SSE/SST$, is included in many of the model selection presentations. In the formula, SSE is for the model being considered and SST is SSE for the single mean model. This is a crude measure of the degree to which the data fits the model. The value R^2 can be misleading in that the model is always at least slightly over-fitted to the data.

Put another way, if the model currently fitted to the data (including the current parameter estimates) were to be used for future predictions, the fit to the new values would almost certainly be lower than the fit to the current data. This is an optimization issue. The parameter values estimated were custom fit to the data in hand, when these values are then used to assess how well this same data fits the model; the fit is just too good. The techniques discussed here are presented as attempts to produce a number like R^2 that indicates how well the current model will fit future data.

11.6.2   Data Splitting

The most obvious approach, if there is a lot of data (often the case with time series), is to split the available data into a training set and a validation set. The model is fitted using the training set, and then R^2 can be computed using the validation set. It is quite obvious that this answers the question: How well does the current model fit a fresh set of data? Picard and Berk (1990) suggest the validation set should always be less than half the data and is usually in the range 1/4 to 1/3 (they offer some complex formulas to optimize this choice, but this is tangential to main themes of this book). A number of complications will be avoided if a period series has a fixed number of whole periods in the training set. These two criteria are sufficient to use cross-validation while avoiding some of the messier formulas required in the general setting (Picard and Cook, 1984; Picard and Berk, 1990).

The New York City temperatures (adjusted for the outlier) data has 14 years of monthly observations. A model will be fitted based on the first 10 years (the training set) and estimation of R^2 for model based on the last 4 years of data (the validation set). The validation set is then 28.6% of the data and the training set is a whole number of periods. It would be possible to use the last 10 years for the training set and the first 4 years for the validation set as well. Other, more creative splits would be possible as well.

The NYC temperature data is scanned and the values are given the name "yNYC." The data is treated as a training and validation set:

```
x_training <- c(1:120)              #create the training set.
y_training <- yNYC[x_training].
x_validate <- c(121:168)            #create the validation.
y_validate <- yNYC[x_validate].
col_2 <- cos(2*pi*x_training/12)    # fit a simple periodic model to the
                                    # training set.
col_3 <- sin(2*pi*x_training/12).
fit <-lm(y_training ~ col_2 + col_3).
```

The fitted model is displayed in Table 11.11.

How well does the data fit the training set and the validation set? The prediction model for the validation set, based on the training set, is

```
validate_fit <- 15.33849 - 2.96592*cos(2*pi*x_validate/12) -2.52051*sin(2*pi*x_validate/12)
```

Figure 11.4 shows a great fit of the model to the data. Table 11.12 indicates the same, the R^2 value found using data splitting is very high in the validation period, and higher than the fit for the training period (a very unusual situation). Why it is unusual for the R^2 value to be higher in the validation period than the training period? In most

TABLE 11.11   The Model Fitted to the Training Set

Coefficients:

	Estimate	SE	t-value	Pr(>|t|)
(Intercept)	15.33849	0.04897	313.24	<2e−16
col_2	−2.96592	0.06925	−42.83	<2e−16
col_3	−2.52051	0.06925	−36.40	<2e−16

Residual SE: 0.5364 on 117 df; multiple R^2: 0.9643; adjusted R^2: 0.9637; F-statistic: 1580 on 2 and 117 df, p-value: <2.2e−16.

Data splitting fit[---]

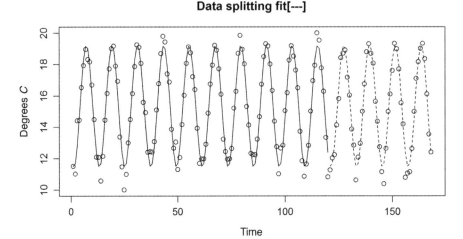

FIGURE 11.4   The model seems to fit future observations quite well.

real world situations, the model is not perfectly stable over time. If the model drifts at all, it is most likely the R^2 value will drop in the validation period due to this drift. It is only when there is no drift in the model that the validation period has a 50–50 chance to be higher in the validation period.

11.6.3   Leave-One-Out Cross-Validation

Not all data sets lend themselves to data splitting. The data set may be too small to split and/or the fitted model may be a local smoother. In the first case, there is too little data upon which to build a model if the data is split; and in the second case, it is not expected the model for any part of the data to directly interpolate/extrapolate to any other part of the model. For these cases, a different approach to cross-validation is possible, something similar to bootstrapping.

Models are fitted to data using an optimization procedure that allows the model to fit the data closely. However, if any particular observation was omitted, and the model fitted to the rest of the data, the model would not fit the omitted data as well as the same model with that observation included. Leave-one-out cross-validation involves systematically excluding each point in the data set and fitting a model with all the

TABLE 11.12   The Value of R^2 for the Training and Validation Sets

Data	SSE	SST	$R^2 = 1 - \text{SSE}/\text{SST}$
Training ($n = 120$)	33.66	942.64	96.43%
Validation ($n = 48$)	11.48	396.27	97.10%

R^2 for the training set can be read directly from the summary(fit) output.

other $n-1$ points. For each observation, the cross-validation residual is the difference between the observation and the model predicted value, when the model value is based on the data with that point excluded. This is like splitting the data into sets of $n-1$ and 1, n times. The result are residuals larger than those under the usual fitted model, and these residuals are used to estimate a new R^2, called R^2_{pred} (a predicted value for R^2 if the model were used to fit a new set of data). These ideas are due to Allen (1971, 1974).

It is not necessary to actually drop each observation and compute a new model; it turns out that linear algebra makes life simple (as usual), and the revised residuals are computed without any extra work on the part of the computer.

There is a number associated with each observation, a diagonal element in a matrix called the hat matrix $[\hat{H} = X(X'X)^{-1}X']$. The number h_{jj}, which is a number between 0 and 1 and happens to be the jth diagonal element of this matrix, is a measure of the leverage of each point. The average value for all h_{jj} is p/n, where p is the number of parameters in the model, and values higher than this represent observations with large leverage (influential points).

The number of interest is computed in the following way (for simple regression models):

$$\text{PRESS} = \sum \left([y_j - \hat{y}_j]/[1 - h_{jj}] \right)^2 \quad \text{and} \quad R^2_{pred} = 1 - \text{PRESS}/\text{SST}.$$

The residuals used in PRESS contrast with the residuals in SSE in that they are larger, and much larger for observations with high leverage. The value R^2_{pred} will often be much smaller than R^2 and will better reflect the degree to which the model would fit a new set of data.

For the NYC temperature (adjusted) data fit, R^2 predicted can be computed as follows:

```
# fit the simple periodic model to all of the data.
time <- c(1:168).
col_2 <- cos(2*pi*time/12).
col_3 <- sin(2*pi*time/12).
fit <-lm(yNYC ~ col_2 + col_3).
var(yNYC)*(168-1)    # find SST              1339.59.
# recover the "hat" values.
# compute the influence measures for the fitted model.
w <- influence(fit).
names(w)      [1] "hat"      "coefficients" "sigma"      "wt.res".
new_res <- fit$residuals/(1- w$hat).
PRESS <- sum(new_res^2).
PRESS            [1] 46.416.
```

$$R^2_{pred} = 1 - 46.416/1339.59 = 96.5\% < R^2 = 96.7\%.$$

Leave-one-out cross-validation produces a result similar to data splitting; the cross-validation R^2 (R^2_{pred}) is not much different from the original R^2 value.

Generally, for time series data, the observations are equally spaced and all of the hat values are the same. In this case, $R^2_{pred} = 1 - \text{SSE}[n/\{n-p\}]/\text{SST}$, so there is no need to even access "hat" values. There will be situations later in the book where this formula is extended to the situations in which hat values cannot be explicitly computed.

11.6.4   AIC as Leave-One-Out Cross-Validation

Akaike's information criterion (Akaike, 1974) is also an attempt to correct for over-fitting bias, but in the likelihood (on the log scale). Akaike views this as the problem of using the same data to both estimate the parameters and estimate the resulting log-likelihood. Because the likelihood in a linear regression model has the sufficient statistic SSE, AIC can also be viewed as a correction for over-fitting when (log)-likelihood maximization is used to estimate SSE. Furthermore, Stone (1977) has shown the asymptotic equivalence of leave-one-out cross-validation (the basis for PRESS) and AIC for model selection.

This idea is discussed in greater detail in Appendix B. However, the implication is that a number can be computed, $\text{PRESS}_{AIC} = \text{SSE} \cdot \exp(2[p+1]/n)$, and an adjustment for cross-validation of the form $R^2_{AIC} = 1 - \text{PRESS}_{AIC}/\text{SST}$ can be computed.

For the NYC temperatures data, $R^2_{AIC} = 1 - 44.77 \cdot \exp(2[4]/168)/1339.59 = 96.5\%$. Interestingly, when comparing any two models when there are no influential observations, $\text{AIC}_1 - \text{AIC}_2 = n \cdot \log_e([1 - R^2_{1,AIC}]/[1 - R^2_{2,AIC}]) \approx n \cdot \log_e([1 - R^2_{1,pred}]/[1 - R^2_{2,pred}]) \approx n \cdot \log_e(\text{PRESS}_1/\text{PRESS}_2)$. This demonstrates a general principle in statistics: when sample size is large, models with only slightly different predictive power have vastly different levels of evidence when compared on a likelihood scale. In other words, the difference in AIC can be large, even when $R^2_{1,pred} \approx R^2_{2,pred}$.

11.7   SUMMARY

This chapter began with hypothesis testing, moved to model selection using information criteria, and finally discussed cross-validation and data splitting.

The general approach of this book is as follows:

(i) In a number of situations, a specific hypothesis is to be tested or a specific confidence interval is to be computed. In these contexts, the usual tools from traditional statistics texts will be used.

(ii) In a number of situations, data will be simulated and a number of models fitted to the data, of which one is the correct model. In these cases, BIC will be used to select the best model.

(iii) In a number of situations, real data will be modeled with no expectation that any of the candidate models is completely correct. In other cases, simulations will be used, but the correct model will NOT be among the models considered. In these cases, the goal is a reliable future prediction. In such cases, data splitting will be considered the gold standard, with AIC and R^2_{pred} used as a substitute for data splitting either because the data set and/or modeling approach makes data splitting infeasible or to save time and focus on other elements of the problem.

In this book, case (iii) is the most common case and R^2_{pred} and AIC the most commonly used tool for model selection. Rarely will R^2_{pred} and AIC select different models.

EXERCISES

1. For the first simulated data (y_1, simple regression), form confidence intervals for the slope and intercept using the summary and assess whether the true intervals contain the true value.

2. For Table 11.6, what would the ANOVA table for a linear fit look like in R?

3. Simulate quadratic data using the previous formulation with the exception that   y_2 <- 7.1 + 2*x_2 + 20*x_2*x_2 + err_2   changed to y_2 <- 7.1 + 2*x_2 + 5*x_2*x_2 + err_2.   In   other   words,   substantially reduce the size of the quadratic term.

    (i) Simulate the data and fit the straight-line and quadratic models.

    (ii) Form a table of information for three models (one mean, a straight-line, and a quadratic fit).

    (iii) Perform the two likelihood ratio tests and state an overall conclusion.

    (iv) Use AIC and BIC to select the best model. Do these results agree with each other and with the conclusion in (iii)?

    (v) Plot both x versus y and the residuals from the straight linear. Do these plots support the conclusion drawn in part (iii)?

4. Simulate quadratic data using the previous formulation with the exception that sigma_2 <- 4.0.
    [If   you   did   exercise   3,   ignore   those   instructions,   use y_2 <- 7.1 + 2*x_2 + 20*x_2*x_2 + err_2 as in the original simulation].

    (i) Simulate the data and fit the straight-line and quadratic models.

    (ii) Form a table of model selection information for three models (one mean, a straight-line, and a quadratic fit).

    (iii) Perform the two likelihood ratio tests and state an overall conclusion.

    (iv) Use AIC and BIC to select the best model. Do these results agree with each other and with the conclusion in (iii)?

    (v) Plot both x versus y and the residuals from the straight linear. Do these plots support the conclusion drawn in part (iii)?

5. Some authors define information criteria as follows:

$$\text{Akaike's information:}\quad \text{AIC} = n \cdot \log_e (\text{SSE}/n) + 2\,(p + 1)$$

$$\text{Schwarz information:}\quad \text{BIC} = n \cdot \log_e (\text{SSE}/n) + p \cdot \log_e(n)$$

Explain why this is consistent with the definitions from the chapter.

6. Compute AIC and BIC for the models from Table 11.7. How much more likely is the quadratic model than the straight-line model in this case (use both AIC and BIC).

7. Show $R^2_{\text{pred}} - R^2_{\text{AIC}} \approx 2 \cdot (1 - \rho^2)/n$ when all the "hat" values are equal. (Where, as sample size increases, $R^2 \to \rho^2$).

8. Justify each equality or approximate equality in the chain:

$$\text{AIC}_1 - \text{AIC}_2 = n \cdot \log_e\left(\left[1 - R^2_{1,\text{AIC}}\right]/\left[1 - R^2_{2,\text{AIC}}\right]\right)$$

$$\approx n \cdot \log_e\left(\left[1 - R^2_{1,\text{pred}}\right]/\left[1 - R^2_{2,\text{pred}}\right]\right)$$

$$= n \cdot \log_e(\text{PRESS}_1/\text{PRESS}_2).$$

9. Using the data "Melbmax.txt" from the exercises folder:

Use data splitting, leave-one-out cross-validation (PRESS), and AIC on this data (remember the data is daily, $f = 365$) to find R^2 values. If your last name begins with A–F, use the last 3 years as the validation set for data splitting; if you name begins with G–M, use the last 2 years; if your last name begins with N–S, use the first 3 years as the validation set; and if your last name begins with T–Z, use the first 2 years as the validation set.

10. (For those familiar with likelihood functions in a general setting).

   (i) Simulate 200 observations from an exponential distribution and compute both the likelihood function and the average of the 200 leave-one-out cross-validation likelihood functions.

   (ii) Perform the entire process in part (i) many times (perhaps 500), get a summary of the difference between the likelihood function and the average of the leave-one-out cross-validation likelihoods over the 500 simulations.

   (iii) Explain how this relates to AIC as a cross-validation value.

12

FITTING FOURIER SERIES

12.1 INTRODUCTION: MORE COMPLEX PERIODIC MODELS

In Chapter 10, a simple periodic model was fitted to the New York City temperature data and a number of more complex data sets were introduced. In the case of more complex models, a model must usually be selected from a collection of candidate models. In this chapter, more complex models are selected that are simple extensions of the previous periodic model, now that model selection tools are available and can address *some* of these issues. In each case, a collection of candidate models are proposed and a "best" model from the candidates is selected. As much as possible, likelihood ratio tests, the Akaike information criterion (AIC), and the Bayesian information criterion (BIC) will be demonstrated, however, given these are all real data sets and the proposed models are probably only crude approximations of reality, BIC is presented only for comparison purposes and will never be used to select the best model.

A second, and perhaps the main, purpose of this chapter will be to introduce an additional complication. Due to this complication, all the results in this chapter should be viewed as tentative. The first fully legitimate model selection will occur in the next chapter, where fitting models to the signal is combined with modeling the error structure.

Basic Data Analysis for Time Series with R, First Edition. DeWayne R. Derryberry.
© 2014 John Wiley & Sons, Inc. Published 2014 by John Wiley & Sons, Inc.

12.2   MORE COMPLEX PERIODIC BEHAVIOR: ACCIDENTAL DEATHS

12.2.1   Fourier Series Structure

Sometimes, data is known, or at least expected, to display periodic behavior, with a known period, but the behavior is more complex than can be captured by the simplest periodic model. For the monthly accidental deaths in the United States (Figure 10.6), it seems clear that the data is periodic, but that the behavior is more complex than the simple periodic model. Richer models are found by fitting additional sine/cosine terms in much the same way as additional terms improve the fit of a Taylor series.

For technical reasons that can be explained in a class covering Fourier series, the manner in which the extension occurs involves the addition of new sine and cosine functions in pairs with a specific structure.

The least-squares problem is $\min \sum (y[j] - h_{\mathrm{pr}}[f \cdot j])^2$, where $f = 1/12$ for monthly data, $f = 1/52$ for weekly data, and $f = 1/365$ for daily data for an annual cycle and the j are any equally spaced time indices. The value pr is the number of cosine/sine pairs:

$$h_1(f \cdot j) = B \cdot \cos(2\pi fj) + C \cdot \sin(2\pi fj)$$

$$h_2(f \cdot j) = h_1(f \cdot j) + D \cdot \cos(4\pi fj) + E \cdot \sin(4\pi fj)$$

$$h_3(f \cdot j) = h_2(f \cdot j) + F \cdot \cos(6\pi fj) + G \cdot \sin(6\pi fj), \quad \text{etc.}$$

12.2.2   R Code for Fitting Large Fourier Series

Some simple R code for fitting any number of trigonometric pairs is presented below.
The strategy is as follows:

Load the data y.
Define the number of pairs desired (pr).
Determine the sample size (n).
Use n and pr to set up a matrix *col_p*.
Using a "for" loop, create the desired columns of sine and cosine pairs.
Fit a model using lm(y ~ col_p).

For model selection purposes $p = 1 + 2 \cdot \mathrm{pr}$ [complexity] and SSE $=$ sum(residuals^2).

Given this information, and the fact that the models are nested, either likelihood ratio tests or AIC can be used to select a "best" or several "good" models. For comparison purposes, a model with a simple mean will be included. The value SSE(= SST) for this model is just var(y)*(n-1).

```
# construction of a matrix
pr <- 5 # number of pairs
n <- length(y)  # sample size
tmp <- rep(NA, 2*pr*n)        #construction of a vector to be reshaped to a matrix
col_p <- matrix(tmp, n, 2*pr)  #reshape the vector tmp to a matrix with
                              # n rows and 2*pr columns
for(k in 1:pr)                # create pr columns of cosines followed by
                              # pr columns of sines
{ for(j in 1:n)
{ col_p[j,k] <- cos(2*pi*j*k/12)
col_p[j, pr+k] <- sin(2*pi*j*k/12)
}}
ones <- rep(1,n)
X <- cbind(ones,col_p)
fit <- lm(y ~ -1+ X)          # for reasons that will become apparent
                              # it makes sense to fit the intercept manually
sum(fit$resid^2)              # print out SSE for the current model
```

12.2.3   Model Selection with AIC

A few comments about Table 12.1 are in order:

(i) Why is the model with six pairs called the "saturated model," and why is the number of parameters 12 instead of 13?

The saturated model is useful in several ways. The saturated model has the smallest possible SSE and usually has the best looking residuals. Residuals from the saturated model will later be used to diagnose the structure of the noise. The notion of a saturated model will be important throughout the book as a modeling tool.

TABLE 12.1   The Model Selection Information for the Accidental Deaths Data with Fourier Series of Various Complexity ($n = 72$)

Model	SSE	P	AIC	BIC	R^2_{pred}
One mean model	65,207,234 (= SST)	1	1299.50	1299.78	NA
One pair	27,578,284	3	1241.54	1246.37	53.9%
Two pairs	21,194,008	5	1226.58	1235.97[a]	62.5%
Three pairs	19,196,147	7	1223.46	1237.39	63.9%
Four pairs	18,244,075	9	1223.79	1242.28	63.5%
Five pairs	15,862,003	11	1217.72[a]	1240.76	66.1%[a]
Six pairs (saturated model)	15,762,389	12	1219.27	1244.59	65.6%

[a]Best model.

A Fourier series will fit a model repeating every 12 months. No matter how many pairs of sine and cosine functions are added. The most complex possible model that can be fitted using this approach has separate 12 means.

(ii) The models display a steady drop in SSE values until the model with five pairs of trigonometric functions. The model with five trigonometric pairs and the saturated model have very similar values for SSE. Often, just by looking at the table, it is obvious which model is the best. Often the SSE values steadily drop with incremental increases in complexity until a best fit is found. When the best fit is found, the SSE values stabilize and each following AIC value increases by about two·(added parameters) units. Referring back to Chapter 11, this was a pattern apparent in many of the model selection tables already constructed.

In this case, the five-pairs model appears to be substantially better than the four-pairs model. Note: $\exp([1223.8 - 1217.72]/2) = 20.9$.

(iii) BIC is included only for comparison. It is the author's belief that BIC would be inappropriate as a model section tool here (Aho et al., 2014). When AIC and BIC differ in model choice, as in this case, BIC selects a less complex model.

12.2.4   Model Selection with Likelihood Ratio Tests

It is also possible to construct a series of likelihood ratio tests to choose between models.

In the best-case scenario, sequential likelihood ratio tests produce a collection of small p-values until the best model is chosen, then all of the p-values are large. In many cases, this does not happen. In this case, the choice of the five-pairs model is clearly the best.

The results of the likelihood ratio tests in Table 12.2 clearly show that three pairs of trigonometric functions are better than fewer pairs, and that five pairs are better than four or six. However, a final test was needed to compare the five- and three-pair models. It seems clear that the five-pair model is the winner. In this context, models are often the null hypothesis for some tests and the alternative hypothesis for others.

To be precise, what can be concluded from these tests are (i) there is strong evidence that at least a five-pair model is required to fit the data and (ii) there is no evidence that six pairs better fit the data than five pairs.

TABLE 12.2   Sequential Likelihood Ratio Tests for the Accidental Deaths Data

Test	χ^2_v	p-value
One pair vs. single mean	61.96 ($v = 2$)	$3.51 \cdot 10^{-14}$
Two pairs vs. one pair	18.96 ($v = 2$)	$7.63 \cdot 10^{-5}$
Three pairs vs. two pairs	7.13 ($v = 2$)	0.0283
Four pairs vs. three pairs	3.66 ($v = 2$)	0.1604
Five pairs vs. four pairs	10.07 ($v = 2$)	0.0065
Six pairs vs. five pairs	0.45 ($v = 1$)	0.5023
Five pairs vs. three pairs	13.74 ($v = 4$)	0.0082

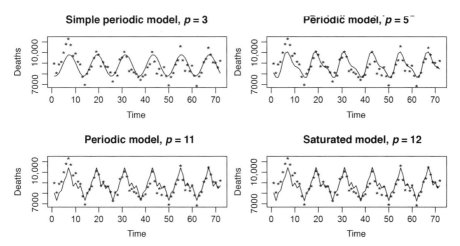

FIGURE 12.1   The fit from a representative set, the various Fourier series.

Figure 12.1 shows how various models fit the data. It seems clear, based on the graphic analysis, that the model chosen by AIC fits the data about as well as the saturated model, but that any model with one or two pairs of trigonometric functions does not capture some of the "jaggedness" clear in the data. In fact, given the shape peaks in the data, it is surprising how well a sum of just a few smooth functions fit the data (Fourier series are amazing—it is good to pause on occasion and remember just how amazing mathematics can be).

12.2.5   Data Splitting

A cross-validation analysis is possible using the first 4 years as the training set, and the last 2 years as the validation set.

Figure 12.2 shows the fit to the last period (dashed line) based on a model using the first 4 years as the training set (solid line).

In this case, the result is quite unusual; the explained variation for the training set is 68.9%, while the same model applied to the validation set produces a value of 85.9%. The model fits the validation set better than the original set (please ignore the fact that the author says this never happens, but it has happened twice already). How can this be? If all of the data show one consistent pattern, there is a 50–50 chance that the explained variation will be higher in the validation set than the training set. Usually, some drift occurs in the model and the fit is better to the training set than the validation set. When the validation set fits better than the training set by a large margin, although this could be due to chance, it is worthwhile to consider outliers or more global anomalous behavior in the training set. Next, a discussion of the residuals and a possible explanation of what is happening in this case are required.

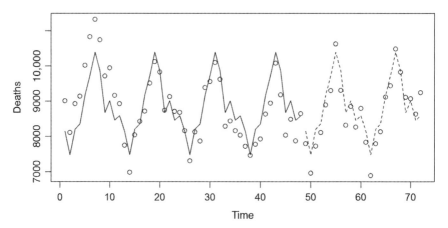

FIGURE 12.2   Training set model (solid line) and validation set model (dashed line) for the accidental deaths data.

12.2.6   Accidental Deaths—Some Comment on Periodic Data

The residuals for the best (five-pairs) model, plotted in temporal order, appear in Figure 12.3. Yikes! What is happening?

For those new to time series, but familiar with regression, the immediate impulse would be to return to the data and fit a quadratic, or some other curvature model. Returning to Figure 12.1, it is obvious there are six distinct peaks and they differ greatly in height. The first peak is much higher than all of the others, while the peaks in the middle are generally low. Because the data is periodic, if a peak is high, all of the points near the peak will usually be high, and if a peak is low, all of the points near the peak will generally be low.

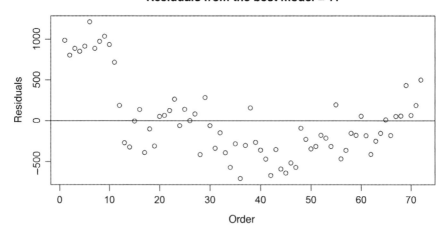

FIGURE 12.3   Curvature in the residual for the fit to the accidental deaths data.

Therefore, it is not really like there are 72 observations supporting a curved feature in the data, it is more like six observations disguised as 72 observations. *Whenever the data is periodic, at some level, there are only as many observations as the number of complete periods.* This global feature of the data suggests caution in understanding more detailed features of the data. While a curvature model might be appropriate for this data, there is too little data to know this, and some skepticism might be in order if such a model were fitted to the data. There are just six peaks and there is a real chance the middle peaks are low and the peaks on the ends are high, just due to random variation. There is often one overriding question—can the model be expected to work well in the future?

At this point, for this data, an impasse has been reached. The residuals do not display a stationary pattern, but a longer series from this data might show this set of residuals to be part of a larger stationary series or might show some sort of trend in the data that must be modeled. Predictions using this model, which always involves extrapolation, should be viewed with caution. It is these anomalies in the data that produce the disparity in fit between the training and validation data.

12.3   THE BOISE RIVER FLOW DATA

12.3.1   The Data

The Boise river flow data is in the Examples folder as "Boise monthly riverflow.txt." For many even moderately large data sets, the data can be overwhelming. It makes sense to consider both all of the data (left side of Figure 12.4) and a subset of the data (right side of Figure 12.4). The global view (all of the data) indicates a relatively stable

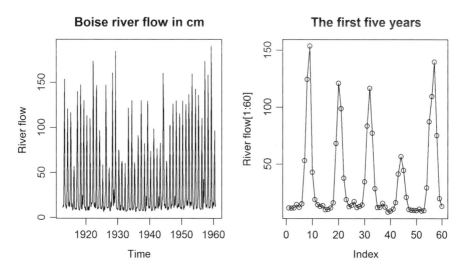

FIGURE 12.4   Two views of the Boise river flow data (DataMarket–Time Series Data Library–Hydrology–Monthly riverflow in cms, Boise river near twin springs, Idaho, October1912–September 1960).

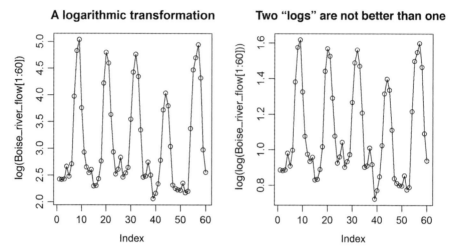

FIGURE 12.5   The transformed data for the first 5 years.

process over time (with peak heights that vary). This can only be ascertained from a global view. On the other hand, the local view (looking at a subset) suggests that a transformation of the data is indicated. This is hard to assess using all of the data.

A logarithmic transformation of the data, as seen in Figure 12.5, reduces the right skew in the data. Since the data still has features indicating right skew (rounded troughs and sharp peaks), a second transformation was used, but the shape did not change further. This data will also be modeled following the usual model selection approach.

12.3.2   Model Selection with AIC

Everything up to now indicates the best model is the one with three periodic pairs (Table 12.3). The model fit seems quite good, both in terms of R^2_{pred} and graphically.

TABLE 12.3   The Model Selection Information for Boise River Flow Data with Fourier Series of Various Complexity ($n = 588$).

Model	SSE	Complexity	AIC	BIC	R^2_{pred}
Single mean model	475.35 (= SST)	1	3628.46	3630.84	NA
One pair	167.83	3	3020.30	3031.43	64.3%
Two pairs	76.94	5	2565.70	2585.58	83.5%
Three pairs	73.83	7	2545.44[a]	2574.07[a]	84.1%[a]
Four pairs	73.76	9	2548.88	2586.27	84.0%
Five pairs	73.55	11	2551.20	2597.35	83.9%
Six pairs (Saturated model)	73.24	12	2550.72	2601.24	83.9%

[a]Best model.

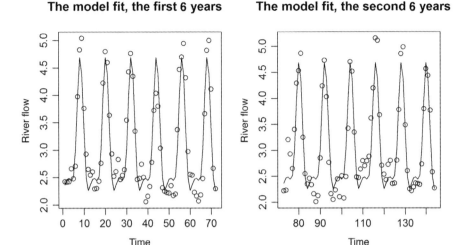

FIGURE 12.6   The fit of the "best" periodic model to the Boise river flow data.

The graph with all 588 observations is too busy, so the model is superimposed on two sets of 6 years of data, of the 49 years of total data (Figure 12.6).

12.3.3   Data Splitting

The requirements that whole years be included in the training set and that 1/4 to 1/3 of the data be in the validation set are quite easy to meet here. The training set should include between 33 and 36 years, leaving between 13 and 16 years for the validation set. In this case, the last 34 years will be used for training, and the first 15 will be used for validation.

The graph of all of the data with the model and validation fit to the model is overly busy and is not presented, but $R^2 = 84.7\%$ for the training set and $R^2 = 83.5\%$ for the validation set.

12.3.4   The Residuals

Of course, there has been a discussion of the 800-pound gorilla in the room. Everything done in Chapters 11 and 12 so far assumes independent errors. It seems quite clear from the *acf()* plots (Figure 12.7), these residuals do not look anything like independent errors. There is hidden benefit in the model so far fitted. Whatever the residuals are, they seem to be about the same for the saturated model and the "best" model.

All the model selection work has been based on a questionable premise, independent errors (reflected by residuals that look like they have no serial correlation). How should AIC, BIC, likelihood ratio tests, and such work in for serially correlated data?

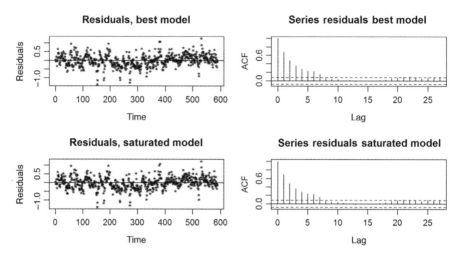

FIGURE 12.7   The residuals with *acf( )* plots. White noise?

12.4   WHERE DO WE GO FROM HERE?

The first step involves making all the right adjustments for data that is AR(1). This will allow for building a foundation in the more complex cases. It should, however be obvious that eventually the model process will involve simultaneously finding a good fit to the signal and a good fit to the noise, and both of these might involve model selection ideas.

EXERCISES

1. Develop the sequential likelihood ratio test for the Boise river flow data. Is the model chosen that same as using AIC?

2. From the Exercises folder scan the file "Furnas 31 78.txt" (aka DataMarket–Time seriesdata library–hydrology–Monthly riverflow in cms, Furnas–vazoes medias mensais, 1931–1978)

   (i) Plot the data. Does it appear periodic? Is a transformation required?

   (ii) Develop a table of model selection information and use AIC to pick the optimal periodic model.

   (iii) Using the same information, develop sequential likelihood ratio tests. What is picked as the best model?

   (iv) Use data splitting (explain how and why you chose the split in the data) to fit a model to a training set and validate the model on the rest of the data. What is R^2 for the training set? The validation set?

   (v) Plot and produce the *acf( )* plot for the residuals from the saturated model and the model picked as best by AIC. How comparable are the residuals?

13

ADJUSTING FOR AR(1) CORRELATION IN COMPLEX MODELS

13.1 INTRODUCTION

Using two cases (a *t*-test and a regression) from the Ramsey and Schafer ("*The Statistical Sleuth*," 2002, Chapter 15), the Semmelweis intervention data (a *t*-test with a different structure than the first case from Ramsey and Schafer), the NYC temperatures (adjusted) data (a simple periodic model), and the Boise river flow data (a complex periodic model), the basic adjustment for AR(1) noise will be presented—it will later be verified that all of these models have residuals indicative of AR(1) models. An approach will be derived for adjusting sample means for AR(1) serial correlation, then a more general approach, often called filtering, is presented, that can be generalized to all AR(m) models in all regression settings. In the first few examples, simulated data will be introduced before the final analysis of the real data. The simulated data can be used to demonstrate how things work out in the best-case scenario—when the true signal is known and the noise is known to be AR(1). In statistics, things can only work out so well, and it is always good to know how much noise there really is in the best-case scenario.

13.2 THE TWO-SAMPLE *t*-TEST—UNCUT AND PATCH-CUT FOREST

13.2.1 The Sleuth Data and the Question of Interest

The first two examples to be discussed come directly from *The Statistical Sleuth* (Ramsey and Schafer, 2002) which has an excellent chapter on AR(1) models,

Basic Data Analysis for Time Series with R, First Edition. DeWayne R. Derryberry.
© 2014 John Wiley & Sons, Inc. Published 2014 by John Wiley & Sons, Inc.

Chapter 15, although this is the only chapter in the book that discusses time series. It should be mentioned that, if you can only own one book on applied statistics, it should be *The Statistical Sleuth*.

Certain logging practices in the Pacific Northwest have been reexamined with a greater concern for the overall environment. Previously, it was common practice to "clear-cut" major portions of a forest, leaving large portions of often steep land bare. One side effect of this practice is that the quality of fish habitat is adversely affected.

Patch-cutting involves checkering forests with sections of clear-cut and undisturbed sections of forest. The first case from Chapter 15 of *The Statistical Sleuth* involves a comparison of nitrate runoff into watersheds in a 5-year period after logging (the measurements were taken about every 3 weeks and there were a total of 88 measurements taken, for our purposes as assumption will be made that the spacing of measurements was equal). One area was undisturbed and the other was patch-cut.

The data is in "SS 151.tx" from the Examples folders, as is read using *read.table()* resulting in the following columns of variables: names(y)

"WEEK" "PATCH" "NOCUT" "y˙P" "y˙N".

The columns y\$PATCH and y\$NOCUT are the data as they appeared in *The Statistical Sleuth* on that data diskette. The columns y\$y_P and y\$y_N are the original data reconstructed by the author and appearing very similar to the Sleuth data as seen in Display 15.1 of that book (Figure 13.1).

Initially, it seems that nitrate levels are higher in the patch-cut region than the uncut region. It is clear that a transformation would improve the general shape of

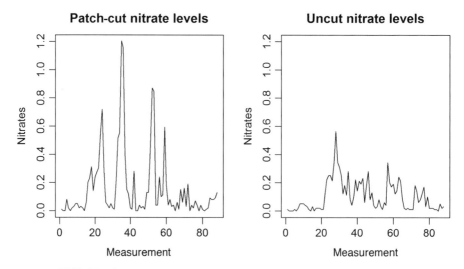

FIGURE 13.1 The nitrate levels for each watershed on the original scale.

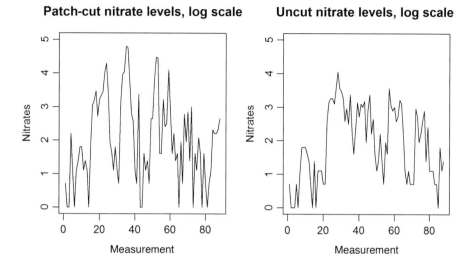

FIGURE 13.2   The nitrate levels after transformation.

the data, but there are zeros. The transformation $w = \log_e(100^*y + 1)$ can be used to rescale and transform the data without producing undefined values (Figure 13.2).

A naïve approach would involve performing a simple *t*-test; this would ignore the serial correlation in the measurements.

The results of this *t*-test give no reason to believe there is any difference in nitrate levels between the two watersheds (Table 13.1). Since there is positive serial correlation between the observations within each sample, it is expected that adjusting for serial correlation will further dilute any apparent differences (this will become obvious mathematically, although it is hoped it is already obvious intuitively). Nevertheless, the details are important.

A simple way to observe the serial correlation is to construct the residuals for this data. In this case, the residuals are sets of data with means removed [the function *acf( )* does this by default]. Based on the *acf( )* plots in Figure 13.3, the two time series have similar, and extremely nonwhite noise, residuals.

TABLE 13.1   Results of a Naïve Comparison of the Nitrate Levels on the Logarithmic Scale

Welch two-sample *t*-test		
$t = 0.6163$,	df $= 169.473$,	*p*-value $= 0.5385$
Alternative hypothesis: true difference in means is not equal to 0		
95 percent confidence interval:		
-0.2423567	0.4623567	
Sample estimates		
Mean of x	Mean of y	
2.018977	1.908977	

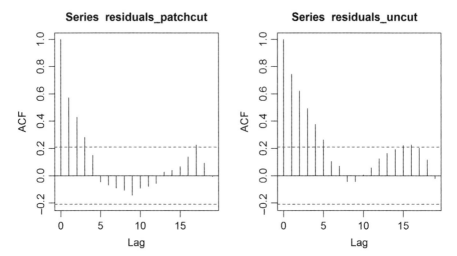

FIGURE 13.3   This is not white noise—serial correlation exists in the residuals.

The function $acf()$ returns $\hat{R}_1 = \hat{a} = 0.573$ for the patch-cut residuals and $\hat{R}_1 = \hat{a} = 0.744$ for the residuals from the uncut watershed. Because the nearby observations are positively correlated, the sample size can be viewed as optimistic (nearby observations contain similar information and are somewhat redundant). There are not $88 + 88$ independent observations. Some kind of adjustment is needed. The first approach to correction for AR(1) serial correlation will involve adjusting the standard errors and then performing the usual two-sample t-test. This allows for easily corrected confidence intervals and yields some useful insights.

13.2.2   A Simple Adjustment for t-Tests When the Residuals Are AR(1)

When the observations are serially correlated, $\mathrm{Var}(\bar{y}) \neq \sigma^2/n$. In a sense, this is the whole problem. For the special case of AR(1) and sample means, it is possible to derive the corrected variance. This will require recalling much of what was done with the AR(1) and AR(2) models earlier.

Recall the autocorrelation function—for lag k, this is $C_k = E(\varepsilon_j \varepsilon_{j\pm k})$, which is the same for all k for a stationary process. This is always the covariance for errors k time units apart. Recall further, $C_0 = \sigma_w^2$, $R_k = C_k/C_0$, and $C_k = C_{-k}$. If residuals, $\hat{\varepsilon}_j$, are used in place of errors, true values are replaced with sample estimates and all these estimates must wear "hats."

What follows employs some formulas from probability that are slight generalizations of the rules for covariance and expectation learned in Chapter 5. In general, covariance is unaffected by a location shift, so, if $y_m = \mu + \varepsilon_m$ and $y_{m+k} = \mu + \varepsilon_{m+k}$, $\mathrm{Cov}(y_{m+k}, y_m) = C_k$.

The more general formula for the variance of a linear combination, when the observations are NOT independent, is as follows.

Let $Y = \sum w_j X_j$. Then $\text{Var}(Y) = \sum w_j^2 \text{Var}(X_j) + 2 \sum_{j<k} w_j w_k \text{Cov}(X_j, X_k)$. In the special case of a sample mean, $w_j = 1/n$ for all j. Furthermore, $E(X_j) = \mu$ and $\text{Var}(X_j) = \sigma_{AR(1)}^2$ for all j. So that

$$\text{Var}(\bar{Y}) = \text{Var}\left(\frac{1}{n}\sum Y_j\right) = \frac{1}{n^2}\text{Var}(Y_1 + \cdots + Y_n)$$

$$= \frac{1}{n^2}\{nC_0 + 2(Cov[Y_1, Y_2] + \cdots + Cov[Y_1, Y_n]) + 2(Cov[Y_2, Y_3] + \cdots +$$

$$Cov[Y_2, Y_n]) + 2(Cov[Y_3, Y_4] + \cdots + Cov[Y_3, Y_n]) + \cdots\}$$

$$= \frac{1}{n^2}\{nC_0 + 2(C_1 + \cdots + C_{n-1}) + 2(C_1 + \cdots + C_{n-2})$$

$$+2(C_1 + \cdots + C_{n-3}) + \cdots\}.$$

Gathering like terms, this becomes

$$\text{Var}(\bar{Y}) = nC_0 + 2(n-1)C_1 + 2(n-2)C_2 + \cdots 4C_{n-2} + 2C_{n-1}$$

$$= \frac{C_0}{n}\left[1 + 2\sum_{j=1}^{n-1}\left(1 - \frac{j}{n}\right)R_k\right].$$

For AR(1), $|a| < 1$, and $R_k = a^k$, $\text{Var}(\bar{Y}) = \frac{C_0}{n}\left[1 + 2\sum_{j=1}^{n-1}\left(1 - \frac{j}{n}\right)a^j\right]$.

If it is further assumed that the sum is dominated by the first several terms, since $R_k \to 0$ for large k, then $\text{Var}(\bar{Y}) \approx \frac{C_0}{n}\left[1 + 2\sum_{j=1}^{n-1}\left(1 - \frac{j}{n}\right)a^j\right] \approx \frac{C_0}{n}\left[1 + 2\sum_{j=1}^{\infty}a^j\right] = \frac{C_0}{n}\frac{1+a}{1-a}$, the last result being the sum of a geometric series. It should be noted this is an approximation and works best when n is large. (Notice the resemblance to the formula from *The Statistical Sleuth*, p. 442).

In other words, the usual variance for any sample mean is adjusted by a factor of $\frac{1+a}{1-a}$. When $a = 0$, for example, which is when the observations are independent, there is no adjustment. When $a > 0$, the usual case in real life, the variance is inflated, (correctly) producing wider confidence intervals and more conservative tests. In the odd circumstance where $a < 0$, the impact is to allow confidence intervals to be more narrow and tests to be more liberal.

13.2.3 A Simulation Example

13.2.3.1 Strategy A routine way to begin understanding a new procedure is to simulate data, apply the procedure, and see if the results are as expected. Two sets of data from an AR(1) model will be simulated. Each sample could have the same or different means, but the same variance and the same value for a. However, each sample will have different white noise. If the procedure works well, the p-values from the procedure will follow a uniform(0,1) distribution when the means are the same and will reject the null hypothesis more when the means are different. Sample sizes and a values are chosen to mimic *The Statistical Sleuth* case, which will be analyzed shortly.

13.2.3.2 The Function arima.sim() in R

A new feature for this simulation is the function *arima.sim()*, which can be used to simulate errors from any AR(m), MA(l) model. At this point help(arima.sim) should be all that the student needs to understand the function. The underlying code for the following simulations:

```
n1 <- 88                  # the same sample sizes as in the first case
n2 <- 88
y1 <- rep(NA,n1)
y2 <- rep(NA,n2)
mu1 <- 10                 # the means can be equal or different
mu2 <- 15                 # here the means are 10 and 15
a <- .66                  # similar to the amount of autocorrelation in the case
noise1 <- arima.sim(n = n1, list(ar=c(a)), sd = 2.0)
noise2 <- arima.sim(n = n2, list(ar=c(a)), sd = 2.0)
y1 <- mu1 + noise1
y2 <- mu2 + noise2
```

13.2.3.3 Simulation Results

The results of 5000 simulations will be collected when the null hypothesis is true (mu1 == mu2) and false. The test-statistic is $z \approx$ $(\bar{y}_1 - \bar{y}_2)/\sqrt{\frac{1+\hat{a}_1}{1-\hat{a}_1}\frac{s_1^2}{n_1} + \frac{1+\hat{a}_2}{1-\hat{a}_2}\frac{s_2^2}{n_2}}$, and, although it might be argued this is a t-statistic, the large sample size indicates that the normal approximation is acceptable, and with the correction for autocorrelation, it is not clear whether Welch's test still applies or how to correct the degrees of freedom otherwise. [It should be noted that there is only one a in an AR(1) model. For the values $\hat{a}_1$ and $\hat{a}_2$ the subscripts refer to samples one and two.]

For these 5000 simulations, information is stored to answer the following questions: (i) How well is the autocorrelation value a approximated? (ii) How do the p-values behave using the estimated values for a (the best we can do)? and (iii) How well do the p-values behave using the true value of a (the absolute best-case scenario)?

Overall, the procedure is working very well (Figure 13.4). Although a is slightly underestimated, on average the first sample had a median of 0.624, with quartiles of 0.562 and 0.680; while the second sample had a median of 0.626, with quartiles of 0.566 and 0.680. The p-values are very close to the uniform(0,1) ideal.

When means are not equal (Figure 13.5), the p-values should concentrate closer to zero, while the estimates of a should still be close to 0.66. The p-values were generally very small. As when the means are equal, the procedure seems to be performing as expected.

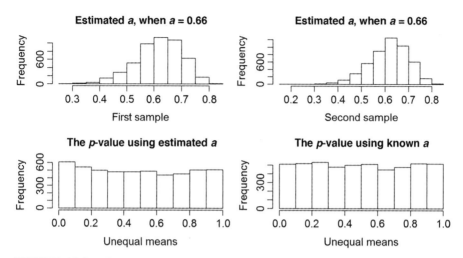

FIGURE 13.4   When the means are equal, the *p*-values should be approximately uniform(0,1), and the estimates of *a* close to 0.66.

13.2.4   Analysis of the Sleuth Data

The first case from *The Statistical Sleuth* involves two different series observed over the same time period. Can it be shown that the mean levels of nitrates are higher in the watershed with runoffs from patch-cut forest, compared to the watershed that has a runoff area that is uncut forest?

Although observational studies have their limitations, observational studies can be designed well or poorly. The student may want to consider, at least in the context of a business, economic, or natural biological environment, why it makes sense to

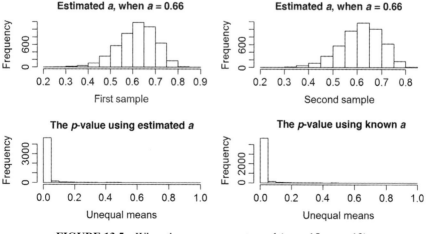

FIGURE 13.5   When the means are not equal ($\mu_1 = 15$, $\mu_2 = 12$).

TABLE 13.2   The Relevant Summary Statistics for the Case

Group	$\bar{y}$	$\hat{a}$	n	$\hat{\sigma}_2$
Patch-cut	2.019	0.573	88	1.631
Uncut	1.909	0.744	88	1.173

observe both time series over the same time span. The student may also want to ponder all of the possible explanations for any differences that are found. If there are differences, are they clearly due to patch-cut versus uncut forest? What could be learned by looking at the watersheds associated with several patch-cut and uncut areas. If there is only one patch-cut area, can anything be learned by looking at this area versus several uncut areas (Exercise 1)?

Given the information in Table 13.2, the test statistic is computed to be:

$$z \approx \frac{\bar{y}_1 - \bar{y}_2}{\sqrt{\frac{1+\hat{a}_1}{1-\hat{a}_1} \frac{s_1^2}{n_1} + \frac{1+\hat{a}_2}{1-\hat{a}_2} \frac{s_2^2}{n_2}}} = \frac{0.11}{\sqrt{0.0683 + 0.0908}} = 0.276.$$

As suspected, there is even less evidence for a difference in means than before the adjustment for serial correlation. Avoid, however, affirming the null hypothesis. This is NOT evidence that patch-cut and uncut forest produce similar nitrate levels in the nearby watersheds. It is evident that any differences in nitrate levels between these two watersheds were too small to detect in this sample. Lots of studies like this could lead to the conclusion that nitrate differences between watersheds in these situations are generally small.

There are really four ways the conclusions of this study could be misunderstood: inappropriately affirming the null hypothesis, overestimating the evidence for causality based on one observational study, overestimating the generalizability of this study as a comparison of all patch-cut versus undistributed sections of forest, and not recognizing that any one study can produce anomalous results (scientific conclusions are based on collections of related studies).

13.3   THE SECOND SLEUTH CASE—GLOBAL WARMING, A SIMPLE REGRESSION

13.3.1   The Data and the Question

The average temperature (°C) for the Northern Hemisphere over periods of 1 year, for the period 1880–1987 (Examples folder, "SS 152.txt"), is presented in Figure 13.6. Is there evidence of a temperature increase for the given period, and if so, can the rate of change be estimated (including a confidence interval)?

Three points are worth noting: the first is that no causal inference follows directly from this study, the slogan "correlation does not equal causation" is just as appropriate

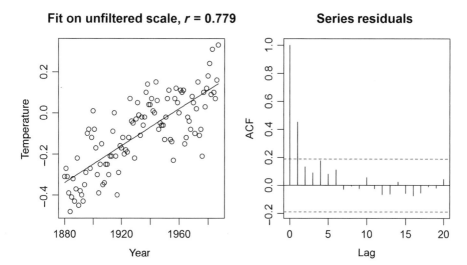

FIGURE 13.6   A line fitted to the data and the *acf( )* plot of the residuals.

here as it is every time a suggestive scatter plot is presented; second, extrapolation is always a concern. Nothing in this plot should be taken as a direct prediction about the future, whatever pattern is found in the data could either reverse or become more pronounced in future years; finally, scientific conclusions are based on the complied results of many studies along with an understanding of the underlying processes that produce the data.

It is clear from Figure 13.6, both that a line fits the data reasonably well, and that the residuals are quite plausibly AR(1). Although the line may very well be a good model for the signal, it is not possible to produce a confidence interval for the slope until an adjustment is made for serial correlation.

13.3.2   Filtering to Produce (Quasi-)Independent Observations

A technique for adjustment of AR(1) discussed previously is insightful, but has limited generalizability. Filtering is a technique that works in a variety of settings and generalizes easily to any AR(m) model. Consider the regression model:

$$y_n = \beta_0 + \beta_1 x_n + \varepsilon_n = \beta_0 + \beta_1 x_n + a\varepsilon_{n-1} + w_n$$
$$y_{n-1} = \beta_0 + \beta_1 x_{n-1} + \varepsilon_{n-1}.$$

Therefore,

$$y_n - ay_{n-1} = \beta_0 (1 - a) + \beta_1 \left(x_n - ax_{n-1} \right) + w_n.$$

In other words, simple linear combinations of the response variables are independent. Although the focus here is on AR(1) models, the general idea is that when given n observations, and an AR(m) model, $n–m$ independent observations can be constructed as linear combinations of the response variables. In this section, filtering will be done with "for" loops. Later, when both the models and/or the order of the AR(m) models increase in complexity, a matrix approach will be presented. Both the "for" loop approach and the approach using a matrix will be useful later; neither is always more practical.

13.3.3 Simulated Example—Regression

A simulation with the known model may lead to a clearer understanding of the modeling approach. The values of n, a, and the line parameters were chosen to closely mimic the case.

```
#set up parameters and constants
time <- c(1:88)
noise <- arima.sim(n = 88, list(ar=c(0.45)), sd = 0.12)   # create and fit the data
y <- -9.0+0.005*time + noise
fit <- lm(y ~ time)
```

In order to construct a filter, the sample estimate of a must be found.

```
z <- acf(fit$resid), the estimated value of a is 0.377.
# construction of a filter, filtered x and y will be length n-1
y_new <- rep(NA,(n-1))
x_new <- y_new
a_hat <- z$acf[2]              # the value of R₁ estimated with acf()
int <- rep(1-a_hat,(n-1))      # the filtered intercept term
# filtering x and y
for(j in 2:n)
{y_new[j-1] <- y[j] - a_hat*y[j-1]
x_new[j-1] <- time[j] - a_hat*time[j-1]}
X <- cbind(int,x_new)
fit_filtered <- lm(y_new ~ -1+ X)   # the intercept is filtered, so a simple intercept
                                    # will not be fitted
```

Three things stand out in Table 13.3: the estimated line itself is not much different in each model, but the standard errors are very much different, in fact the standard errors are larger in the filtered model, indicating that the correct confidence intervals will be wider than the confidence intervals that ignore the AR(1) structure in the residuals. Recall, in Chapter 5, it was learned $\sigma^2_{AR(1)}(1 - a^2) = \sigma^2_w$, hopefully it is not then surprising that

$$0.1461^2(1 - 0.377^2) \approx 0.1336^2.$$

Because the confidence intervals are wider for the filtered model, it is not surprising that the level of fit of the model to the data is lower. It is clear from Figure 13.7 that

TABLE 13.3   A Straight Line, AR(1) Model with and without Filtering

The original data coefficients:

	Estimate	SE	t-value	Pr(>|t|)
(Intercept)	−8.979239	0.031409	−285.883	< 2e−16
Time	0.004836	0.000613	7.889	8.88e−12

Residual SE: 0.1461 on 86 df

The filtered data coefficients:

	Estimate	SE	t-value	Pr(>|t|)
Xint	−9.004357	0.053197	−169.26	< 2e−16
Xx_new	0.005515	0.001242	4.44	3.34e−05

Residual SE: 0.1336 on 69 df

the filtering has reduced the correlation between the variables but produces residuals that look a lot more like white noise [*acf( )* plots].

13.3.4   Analysis of the Regression Case

The global warming data, after fitting a line, has an estimated value of $\hat{a} \approx 0.452$. This is the value used in filtering.

With the filtered model (Table 13.4) it is possible to construct a confidence interval for the slope: $0.0046033 \pm 2 \cdot 0.0005804 = (0.00344, 0.005764)$. In other words, the rate of change in temperature is about 0.34–0.58°C per century. Given the very small *p*-value for the slope, there is little questioning of the fact that average temperatures increased between 1800 and 1987. However, as previously stated, this study has the usual caveats associated with any single statistical analysis.

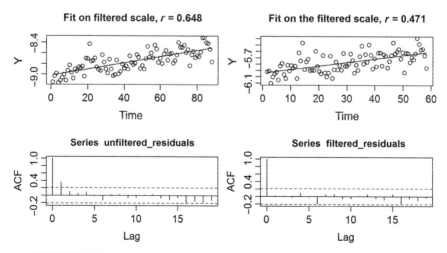

FIGURE 13.7   A comparison of the filtered and unfiltered fit for simulated data.

TABLE 13.4   The Global Warming Data with and without AR(1) Filtering

The original fit coefficients:

	Estimate	SE	*t*-value	Pr(>|*t*|)
(Intercept)	−8.7867143	0.6795784	−12.93	<2e−16
Year	0.0044936	0.0003514	12.79	<2e−16
Residual SE: 0.1139 on 106 df				

The filtered fit coefficients:

	Estimate	SE	*t*-value	Pr(>|*t*|)
Xintercept	−8.9980900	1.1231685	−8.011	1.65e−12
Xnew_year	0.0046033	0.0005804	7.931	2.48e−12
Residual SE: 0.1016 on 105 df				

The relationship on the filtered scale (Figure 13.8) is weaker ($r = 0.612$ versus $r = 0.779$), but with observations that appear more like independent observations (the residuals are closer to white noise).

13.3.5   The Filtering Approach for the Logging Case

Let the transformed patch-cut numbers be denoted "Patch" and the transformed nitrate numbers from the uncut watershed be denoted "Uncut." In this case, the models are (for each sample) $y_m - ay_{m-1} = \mu(1-a) + w_m$. The estimated means are therefore based on the transformed observations: $(y_m - ay_{m-1})/(1-a) = \mu + w_m/(1-a)$.

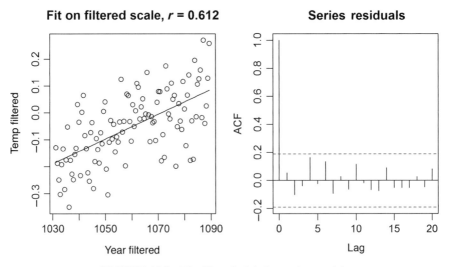

FIGURE 13.8   The filtered global warming model.

TABLE 13.5   Using Filtered Values to Perform a t-Test

Welch two-sample t-test		
Data:	new_patch and new_uncut	
$t = 0.2971$,	df $= 169.003$,	p-value $= 0.7668$
Sample estimates:		
Mean of x	Mean of y	
2.064061	1.945923	

The filtering then is as follows:

```
n <- 88
new_Patch  <- rep(NA,n-1)
new_Uncut  <- rep(NA, n-1)
for(j in 1:(n-1))
{ new_Patch[j] <- Patch[j+1]-0.573*Patch[j]      # recall the estimates of a were
new_Uncut[j] <- Uncut[j+1] - 0.744*Uncut[j]}     # 0.573 and 0.744
new_Patch <- new_Patch/(1-0.573)                 #filtering followed by this step
new_Uncut <- new_Uncut/(1-.744)                  # produces (y_m − ay_{m−1})/(1 − a)
```

Now, a simple t.test() gets the correct result displayed in Table 13.5:

```
t.test(new_Patch,new_Uncut)
```

The p-value from this test is very similar to the previous p-value.

13.3.6   A Few Comments on Filtering

It must always be remembered that statistics deals in estimates. The relationship $y_m - ay_{m-1} = \mu(1 - a) + w_m$ is true, but, in fact the adjustment implemented was $y_m - \hat{a}y_{m-1} = \hat{\mu}(1 - \hat{a}) + \hat{w}_m$. The filtering procedure does not, in practice, produce independent observations. The observations produced are only approximately independent; they will be called quasi-independent throughout the rest of the book to remind the reader that filtering is always imperfect. The quality of the filtering depends on the condition $\hat{a} \approx a$ and will later depend on $\hat{a}_1 \approx a_1 \ldots \hat{a}_m \approx a_m$. This highlights the fact that large sample sizes are desired for reliable time series analysis.

More generally, for almost all statistical procedures, there are sources of uncertainty not included in the standard error. Here no explicit adjustment has been made for replacing the true value of a with a sample estimate. To the extent that any confidence interval has not captured every possible source of uncertainty, that interval is actually narrower than a perfect interval; this implies that all p-values are smaller than a perfect p-value (exercises).

Finally, it should be noted that the quick-and-dirty calculation $\sqrt{1 + \hat{a}}/\sqrt{1 - \hat{a}}$ often produces a ballpark estimate of how much any standard errors will be affected when AR(1) is used to model serial correlation.

13.4   THE SEMMELWEIS INTERVENTION

13.4.1   The Data

The following was taken directly from Wikipedia. Although citing an online source is a bit unusual, the story is well known. The summary presented here is accurate and freely accessible:

> Ignaz Philipp Semmelweis[Note 1] (July 1, 1818–August 13, 1865) was a Hungarian physician now known as an early pioneer of antiseptic procedures. Described as the "savior of mothers",[1] Semmelweis discovered that the incidence of puerperal fever could be drastically cut by the use of hand disinfection in obstetrical clinics.[1] Puerperal fever was common in mid-19th-century hospitals and often fatal, with mortality at 10%–35%. Semmelweis postulated the theory of washing with chlorinated lime solutions in 1847[1] while working in Vienna General Hospital's First Obstetrical Clinic, where doctors' wards had three times the mortality of midwives' wards. He published a book of his findings in *Etiology, Concept and Prophylaxis of Childbed Fever*.
>
> Despite various publications of results where hand-washing reduced mortality to below 1%, Semmelweis's observations conflicted with the established scientific and medical opinions of the time and his ideas were rejected by the medical community. Some doctors were offended at the suggestion that they should wash their hands and Semmelweis could offer no acceptable scientific explanation for his findings. Semmelweis's practice earned widespread acceptance only years after his death, when Louis Pasteur confirmed the germ theory and Joseph Lister practised and operated, using hygienic methods, with great success. In 1865, Semmelweis was committed to an asylum, where he died, ironically, of septicemia at age 47.

For 36 months Semmelweis kept track of the infant mortality at his hospital, where many infant deliveries occurred each month. For the first 17 months no intervention occurred; for the last 19 the surgical instruments were sterilized between surgeries.

Analysis of the Semmelweis data (Broemeling, 2013) will now be relatively straightforward. The analysis will require filtering, and some care must be taken to get the underlying structure just right for this filtering.

The Semmelweis data ("Semmelweis.txt" from the Examples folder) should be read using *read.table( )* with header=TRUE. The columns are: names(assigned name)

"intervention" "births" "deaths" "mortality".

The impact of the intervention seems quite obvious (Figure 13.9), however, the gains from intervention can be quantified. The right panel in Figure 13.9 is for the transformed data.

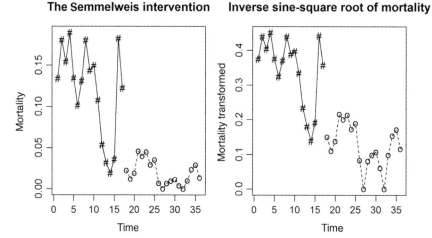

FIGURE 13.9   Mortality in the nonintervention period (first 17 observations—#) and intervention period (last 19 observations—o).

The main objectives here are to determine the degree of evidence that the intervention was effective and to estimate (including confidence intervals and *p*-values) the impact of the interventions (is infant mortality reduced a little or a lot?).

13.4.2   Why Serial Correlation?

It will become clear that the data has serial correlation, but what is the source of the correlation? One skill expected of any experienced statistician is that they know when to expect serial correlation and when not to, and to be able to convince others that serial correlation could be an issue, when the problem could be present.

Putting the question a bit more concretely, why would someone expect infant mortality rates to be more similar from one month to the next than those measured several months apart? In fact, there are several aspects of any hospital that change slowly over time. There are seasonal illnesses, and there is turnover in medical staff. These factors both influence mortality rates and may be constant from one month to the next, but vary greatly over much larger time frames. Here are probably many other such factors that others more familiar with hospitals could imagine.

13.4.3   How This Data Differs from the Patch/Uncut Case

This analysis, like the first case in this chapter, will apparently involve a two-sample *t*-test. Nevertheless, the structures of the studies are very different. In the logging study, two series were observed over the same period of time. In this study, one time series is observed over a period of time, but, at some point, an intervention occurs and there is an interest in whether the intervention made a difference. An interesting question (an exercise)—describe how the logging case would be carried out as an intervention and describe how the Semmelweis case would have been carried out in a fashion similar to the logging case. Further, discuss some threats to inferring cause–effect associated with each design.

13.4.4 Filtered Analysis

Before any analysis can begin, the data must be transformed. The observations are proportions. There are two commonly used transformation for such data, $y = \sin^{-1}(\sqrt{p})$ and $y = \log_e(p/[1-p])$. The former transformation will be used in this analysis and the latter is left as an exercise. The R code modified_m <- asin(sqrt(mortality)) was used to produce the plot on the right panel of Figure 13.9.

To understand the filtering procedure, begin by thinking about the matrix representation of the data, letting μ_I be the mean for the intervention period, and letting μ_N be the mean for the nonintervention period (there are several ways to parameterize this model, but the parameterization chosen is best for producing individual confidence intervals for μ_I and μ_N):

$$\begin{pmatrix} y_n \\ \cdots \\ y_1 \end{pmatrix} = \begin{bmatrix} 1 & 0 \\ 1 & 0 \\ \cdots & \cdots \\ 0 & 1 \end{bmatrix} \begin{pmatrix} \mu_I \\ \mu_N \end{pmatrix}.$$

The filtering is exactly the same as before, treating x_j as column $j = 1,2$: $y_n - ay_{n-1}$, and $x_{j,n} - ax_{j,n-1}$. Before filtering the data, the unfiltered model will be fitted for comparison purposes and to examine the residuals (producing $\hat{a}$).

```
x_1 <- c(rep(1,17),rep(0,19))
x_2 <- c(rep(0,17), rep(1,19))
X <- cbind(x_1,x_2)
fit <- lm(modified_m ~ -1 + X)
```

It is clear in Figure 13.10 that when a simple mean is fitted to each group, the residuals do not look like white noise. From the *acf()* function applied to the residuals it is determined that $\hat{a} = 0.6149$.

The data can be filtered, as before, with a "for" loop.

```
new_x1 <- rep(NA,35)
new_x2 <- rep(NA,35)
new_y <- rep(NA, 35)
for(j in 1:35)
{ new_x1[j] <- x_1[j+1] - 0.6149*x_1[j]
new_x2[j] <- x_2[j+1] - 0.6149*x_2[j]
new_y[j] <- modified_m[j+1] - 0.619*modified_m[j]}
X <- cbind(new_x1, new_x2)
fit_filtered <- lm(new_y ~ -1 + X)
```

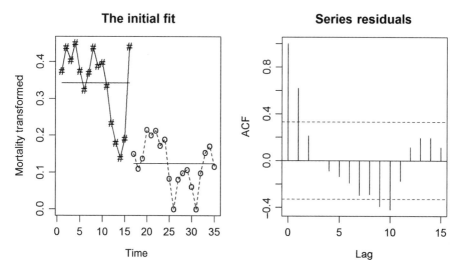

FIGURE 13.10   The fit to the transformed data before filtering.

When filtering the data (Table 13.6) the R^2 value produced in the R output is not the appropriate value (the issue is that a no-intercept model is being fitted). Using the general formula $R^2 = 1 - SSE/SST$ and the model residuals, it was possible to compute R^2 for each model.

Is there a difference between the intervention and nonintervention period?

$z \approx (0.33336 - 0.12251)/\sqrt{0.04010^2 + 0.03472^2} = 3.98$; there is convincing evidence of a difference in median mortality between the periods.

The nonintervention period had a median mortality rate of 6.3–16.2% (0.25316–0.41356, inverted with $\sin[c]^2$, remember the sine function requires radian measure). The intervention period had a median mortality rate of 0.3–3.6%. Overall, the sterilization process has almost certainly (statistics never allows certainty) reduced mortality rates. Of course, from a modern perspective, now fully embracing the germ theory,

TABLE 13.6   The Results for the Filtered and Unfiltered Models Compared

The unfiltered model coefficients:

	Estimate	SE	t-value	$Pr(>|t|)$
Xx_1	0.34361	0.01985	17.307	$< 2e{-}16$
Xx_2	0.12406	0.01878	6.606	$1.42e{-}07$

Residual SE: 0.08186 on 34 df ($R^2 = 65.5\,\%$)

The filtered model coefficients:

	Estimate	SE	t-value	$Pr(>|t|)$
Xnew_x1	0.33336	0.04010	8.313	$1.33e{-}09$
Xnew_x2	0.12251	0.03472	3.529	0.00125

Residual SE: 0.06525 on 33 df ($R^2 = 37.2\,\%$)

these results are not at all surprising. Remember, hindsight, especially hindsight 150 years hence, is always 20–20.

13.4.5   Transformations and Inference

The use of transformations is routine both in this book and in statistical practice. However, this is the first time, in this book, that sufficient theory has been developed so that inferences (hypothesis tests and confidence intervals) are possible.

Transformations alter inference. Most inferential procedures are based on means or weighted means (in the case of regression). However, means do not respect transformations. That is, the mean of transformed data is almost never the same as the transformation of the mean. For example, in the example just completed, the transformation used was $y = \sin^{-1}(\sqrt{p})$. Suppose five values are given for p: 0.11, 0.21, 0.09, 0.08, and 0.04. If the sample values are transformed, they are 0.33807, 0.47603, 0.30469, 0.28676, and 0.20136. The mean of these numbers is 0.321. When working on the transformed scale, this is the sample mean. On the other hand, the mean of the original numbers is 0.106 and the transformation of the sample mean is $\sin^{-1}(\sqrt{0.106}) = 0.332$.

On the other hand, the median does respect (monotone) transformations. For the five values given the median is 0.09 and $\sin^{-1}(0.09) = 0.30469$ is the median of the transformed data. Inference for transformed data is based on the following logic:

- If the transformation is moderately successful, the resulting sample values follow (at least approximately) a normal distribution.
- For a normal distribution, the median and the mean are about equal.
- For a normal distribution, the confidence interval for the mean is therefore also a confidence interval for the median.
- The interval, understood as an interval for the median, can be back-transformed (using the function inverse of the original transformation) to produce an interval for the median on the original scale.

This logic applies to all monotone transformations for sample data; however, there are certain specialized results for the logarithmic transformation. *The Statistical Sleuth* (Ramsey and Schafer, 2002) discusses these specialized results in the case of the two-sample *t*-test (Section 3.5) and simple regression (Section 8.4). Much more care is taken in the discussion of transformations in *The Statistical Sleuth* than in any other book of which the author is aware.

13.5   THE NYC TEMPERATURES (ADJUSTED)

13.5.1   The Data and Prediction Intervals

In the previous work, a simple periodic model for the NYC temperatures was found to be quite effective. Forecasters might be interested in predicting the monthly average temperature for the first few months of 1960 (the time series ended in December 1959). On the other hand, the modeler might be interested in making prediction intervals and checking their reliability. Therefore, a model will be fitted to the first 165 observations, a prediction interval for observations 166–168 will be made, and the results will be compared to the actual observations in those periods. A model for AR(1) prediction intervals will be developed and then applied to this data. The data, with the outlier corrected, appears in Figure 10.3.

In time series analysis, it is often the case that future forecasts are required. Three elements are introduced with this new inferential procedure: extrapolation, prediction intervals, and greater reliance on assumptions.

Prediction about the future assumes that the statistical model will continue to fit future data. There are several reasons this is often implausible, but it also seems clear that the model will often degenerate slowly in quality, so that the model will fit data only a few periods in the future almost as well as the data used to fit the model. To some degree, the reliability of extrapolation into the future involves subject-matter expertise. In other words, when modeling the sunspot data, an astrophysicist would likely know something about the long-run stability of the current model. When dealing with the economy, a macroeconomist might know how long a current model will hold, and what might cause it to fail.

Once the decision has been made that a model might continue to fit the data for a period in which predictions are to be made, a prediction interval is possible. A confidence interval is for a mean or median, but a prediction interval is for a single observation. For example, suppose a sample mean is computed based on n observations. A confidence interval for the true mean is $\bar{y} \pm ts/\sqrt{n}$. On the other hand, suppose an additional observation is found, how might this additional observation behave? *If the data is normal*, the interval is valid. It is much wider than a confidence interval; being $\bar{y} \pm ts\sqrt{(n+1)/n}$ (Devore, 2008, pp. 274–275).

A key third feature is that an assumption of normality is required. For a confidence interval, the central limit theorem plays a role in the reliability of the interval because the sample mean is often approximately normal even when the underlying data is not. A prediction interval has no such protection. The shape of the interval reflects the shape of the underlying distribution. It is more important to examine carefully the normality assumption by checking the residuals (there are approaches to prediction intervals that are less dependent on assumptions, which will not be explored here).

No matter how good the residuals look, the underlying distribution that produced the residuals may not be normal and the underlying statistical model will almost certainly not hold for future prediction as well as for the current data. Furthermore, the prediction intervals produced here will usually replace a with $\hat{a}$ without including the additional uncertainty that comes from replacing a parameter with an estimate. The prediction intervals presented, for the same reasons as mentioned before, have not incorporated all possible sources of uncertainty and are (perhaps only a bit) too narrow.

For all their limitations, prediction intervals are probably reported too little. It is important to acknowledge variability as well as averages. Nate Silver's book, *The Signal and the Noise*, has many examples of the problems of reporting only estimates without including any measures of uncertainty (in weather forecasting, elections, etc.). This unreported variability is often the large variability from time to time (prediction intervals), not the smaller variability in averages (confidence intervals).

13.5.2   The AR(1) Prediction Model

The AR(1) structure is on the error (the noise). Recall that the basic structure is $\varepsilon_j = a\varepsilon_{j-1} + w_j$. Suppose a prediction is to be made in period $j+1$, and the values ε_m, for $m = 1 \ldots j$, are known, $m > j$ are unknown. Then $E(\varepsilon_{j+1}) = a\varepsilon_j$ and $\mathrm{Var}(\varepsilon_{j+1}) = \sigma_w^2$, continuing with

$$\varepsilon_{j+2} = a\varepsilon_{j+1} + w_{j+2} = a^2\varepsilon_j + aw_{j+1} + w_{j+2},$$
$$\varepsilon_{j+3} = a\varepsilon_{j+2} + w_{j+3} = a^3\varepsilon_j + a^2 w_{j+1} + aw_{j+2} + w_{j+3}.$$

So $E(\varepsilon_{j+2}) = a^2\varepsilon_j$ and $\mathrm{Var}(\varepsilon_{j+2}) = (1 + a^2)\sigma_w^2$, while $E(\varepsilon_{j+3}) = a^3\varepsilon_j$ and $\mathrm{Var}(\varepsilon_{j+3}) = (1 + a^2 + a^4)\sigma_w^2$. In general, $E(\varepsilon_{j+k}) = a^k\varepsilon_j$ and $\mathrm{Var}(\varepsilon_{j+k}) = \sum_{l=0}^{k-1} a^{2l}\sigma_w^2$ for $k = 1,2,3\ldots$. So $E(\varepsilon_\infty) = 0$ and $\mathrm{Var}(\varepsilon_\infty) = \sigma_w^2/(1 - a^2) = \sigma_{AR(1)}^2$.

Of course, the errors up to any point are never known, but the residuals, $\hat{\varepsilon}_j$, are known for $j = 1 \ldots n$. From the sample, the values $\hat{a}$ and $\hat{\sigma}_{AR(1)}^2$ are also known, which allows for the computation of $\hat{\sigma}_w^2$.

13.5.3   A Simulation to Evaluate These Formulas

The next step is to simulate AR(1) errors and assess the success of the formulas outlined above. One key concern is that all of $\hat{\varepsilon}_j, j = 1 \ldots n$, $\hat{a}$, and $\hat{\sigma}_{AR(1)}^2$ are estimates, whereas the formulas were based on known values. The simulation is straightforward:

(i) Simulate n AR(1) errors ($n = 200$, $a_1 = 0.7$, $\sigma_w = 2.0$, implying $\sigma_{AR}^2 = 7.84$).
(ii) Compute the residuals from the data.
(iii) Use the residuals to compute $\hat{a}$ and $\hat{\sigma}_{AR(1)}^2$.

This simulation produces the data in Figure 13.11.

The simulated residuals are obviously AR(1) in nature with positive correlation; the residuals cluster together with long sequences of residuals tracking in local clusters. For this data $\hat{a} = 0.767$ and $\hat{\sigma}_{AR(1)}^2 = 9.948839$. The prediction intervals in the graph $\pm 2\hat{\sigma}_{AR(1)}$ appear to be about right, with about 95% of the observations falling inside the prediction bands.

However, using the details of the AR(1) structure, it is possible to produce much narrower prediction intervals for these same residuals.

Both of the graphs shown in Figure 13.12 are based on this same set of residuals.
Graph 1 (Figure 13.12, left panel)—predictions, one step ahead.
Fit predictions to all of the observations based on the formula $\hat{\varepsilon}_{j+1} = \hat{a}\hat{\varepsilon}_j$, where the jth residual is used to predict the $j+1$ residual.

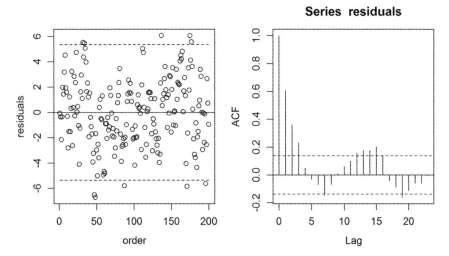

FIGURE 13.11   The residuals from simulated data with a clear positive AR(1) structure.

Apply prediction bands by forming the relationship $\hat{a}\hat{\varepsilon}_j \pm 2\hat{\sigma}_w$.

Graph 2 (Figure 13.12, right panel)—predictions, two steps head.

A second graph will be used to further verify the iterative nature of the formulas.

Fit prediction to all of the observations based on the formula $\hat{\varepsilon}_{j+2} = \hat{a}^2\hat{\varepsilon}_j$, where the jth residual is used to predict the $j+2$ residual.

Apply prediction bands by forming the relationship $\hat{a}^2\hat{\varepsilon}_j \pm 2(1 + \hat{a}^2)\hat{\sigma}_w$.

The prediction intervals are all of different widths (Figures 13.11 and 13.12). All perform correctly, containing about 95% of the data. The ones in Figure 13.11

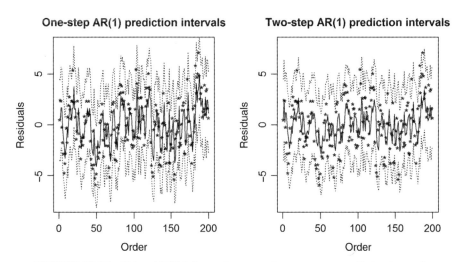

FIGURE 13.12   Using AR(1) information to produce narrower prediction intervals.

TABLE 13.7   Filtering the Model: $\hat{a} = 0.129$ and $\hat{\sigma}_{AR(1)} = 0.5244$

The unfiltered model coefficients:

	Estimate	SE	t-value	Pr(>|t|)
(Intercept)	15.29996	0.04084	374.64	<2e−16
Xcol_c	−2.99295	0.05793	−51.66	<2e−16
Xcol_s	−2.53916	0.05757	−44.10	<2e−16
Residual SE: 0.5244 on 162 df				

The filtered model coefficients:

	Estimate	SE	t-value	Pr(>|t|)
new_Xintercept	15.29994	0.04678	327.04	<2e−16
new_Xnew_c	−2.99321	0.06502	−46.03	<2e−16
new_Xnew_s	−2.54040	0.06436	−39.47	<2e−16
Residual SE: 0.5216 on 161 df				

enlist no information about the AR(1) structure of the residuals and are the widest $(\pm 2\hat{\sigma}_{AR(1)})$. The one-step prediction intervals (Figure 13.12, left panel) use the AR(1) information only one step beyond the data; these intervals have width $\pm 2\hat{\sigma}_w = \pm 2\sqrt{(1 - \hat{a}^2)}\hat{\sigma}_{AR(1)}$, while the two-step interval (Figure 13.12, right panel), which uses AR(1) information but involves a step further from the known data, has width $\pm 2\sqrt{(1 + \hat{a}^2)}\hat{\sigma}_w = \pm 2\sqrt{(1 - \hat{a}^4)}\hat{\sigma}_{AR(1)}$. The more steps taken in the future, the wider will the prediction intervals become. At the limit (an infinite number of steps in the future) any value of knowing the errors from AR(1) is lost.

13.5.4   Application to NYC Data

As promised, a model will be fitted to observations 1–165, and prediction intervals will be produced for observations 166–168. In this manner, an informal evaluation of the performance of the intervals is possible. In the spirit of doing everything right, the model will be filtered, although this will change the prediction model very little (filtering is critical for correct estimates of the standard errors and inference, but often has very little impact on prediction—see exercises). The R code for the filtered model is left as an exercise.

Obviously, the impact of filtering is minor for this data (Table 13.7). Because $\hat{a} = 0.129$, the residual error and standard errors for both models are quite similar. Recall that, for large samples, the point estimates will usually be quite similar for the filtered and unfiltered model, but the standard errors will usually be larger for the filtered model, but the degree of change will depend on the level of dependence (serial correlation) between observations.

Prediction intervals (Table 13.8) capture the natural variability in the data. Obviously, from this data, it appears that even monthly averages vary a lot from year to year. In reading Table 13.8, consider observation 166. The predicted value from the signal is 16.002, however, the residual from observation 165 is +0.507. In an AR(1)

TABLE 13.8   Predictions for the Next Three Periods, Based on the Periodic Nature of the Data. The Last Residual for the Fitted Model, $\hat{\varepsilon}_{165}$, Is 0.5068881 and $\hat{\sigma}_w = \sqrt{1 - \hat{a}^2}\, \hat{\sigma}_{AR(1)} = 0.5200$

Observation	Observed value	Lower bound	Prediction $\hat{y} + \hat{a}^k \varepsilon_{165}, k = 1,2,3$	Upper bound
166	16.115	$16.067 - 2(0.52)$ $= 15.027$	$16.002 + 0.129(0.507)$ $= 16.067$	$16.067 + 2(0.52)$ $= 17.107$
167	13.352	$13.986 - 2(0.5243)$ $= 12.937$	$13.978 + 0.129^2(0.507)$ $= 13.986$	$13.986 + 2(0.5243)$ $= 15.035$
168	12.398	$12.308 - 2(0.5244)$ $= 11.259$	$12.307 + 0.129^3(0.507)$ $= 12.308$	$12.308 + 2(0.5244)$ $= 13.356$

model, this indicates that the errors are trending above the signal. The estimate is that the next residual will be $\hat{a}\hat{\varepsilon}_{165} = 0.129\,(0.507) = 0.0654$, so the prediction is 16.0674. The observed value was 16.115.

The prediction intervals slowly become wider with each step, but are already near the final width. These intervals are narrower than other intervals computed by, for example, software, where additional uncertainty related to replacing parameters with estimates is included in the interval. Nevertheless, for large samples, the intervals are essentially the same as these.

13.6   THE BOISE RIVER FLOW DATA: MODEL SELECTION WITH FILTERING

13.6.1   The Revised Model Selection Problem

The Boise river flow data ("Boise monthly riverflow.txt," Examples folder, Figure 12.5) will be used to put the various elements of model selection together. The data will be fitted using, as before, for transformed data $y' = \log_e(y)$.

A model for the data will be selected from among the various periodic models (trigonometric pairs $= 0, 1, 2, 3, 4, 5$, or 6) to fit the data. The procedure is as follows:

Fit a saturated model.

Use the residuals from the saturated model to find $\hat{a}$.

Fit filtered models for each pair (on this scale the observations are quasi-independent).

Perform model selection (AIC, likelihood ratio tests, or R^2_{pred}) on the filtered models.

13.6.2   Comments on R^2 and R^2_{pred}

Before fitting the models, an important issue must be addressed. For filtered models, there are two values for R^2 and R^2_{pred}, the value for the original data and the value for the filtered data (denoted R_{F}^2 and $R_{\mathrm{F,pred}}^2$). How are these two values related, which is "correct" for model selection, and which best reflects the predictive accuracy of the model?

Model selection tools assume independent observations. So model selection must be performed on the filtered scale. However, for the real world, the filtered scale is an artificial mathematical construct. The quality of fit of any model chosen should be reported on the original scale.

For the unfiltered model, it can be shown (exercise) that $R^2 = \dfrac{\hat{\beta}_1^2 \mathrm{Var}(x)}{\hat{\beta}_1^2 \mathrm{Var}(x) + \hat{\sigma}_{\mathrm{AR}(1)}^2}$ and that for the filtered model $R_{\mathrm{F}}^2 = \dfrac{\hat{\beta}_1^2 \mathrm{Var}(x)}{\hat{\beta}_1^2 \mathrm{Var}(x) + \frac{1+\hat{a}}{1-\hat{a}} \hat{\sigma}_{\mathrm{AR}(1)}^2}$ (exercise). In this context $\hat{\beta}_1^2$ is the estimated slope for the regression squared, $\mathrm{Var}(x)$ is the sample variance of the x values, and $\hat{\sigma}_{\mathrm{AR}(1)}^2$ is the sum of the squared residuals from the model. So it is expected that, for most real problems, $R_{\mathrm{F}}^2 < R^2$.

Since it is possible to directly compute $R_{\mathrm{F, pred}}^2$ on the filtered scale, the computation of R_{pred}^2 does present some issues. The value for the original scale will be computed in the following manner.

The point estimates from the filtered scale can be used to compute residuals on the original scale. The "hat" values from fitting a model on the original scale can be used to determine their influence. Then *PRESS*-like residuals can be formed by dividing the residuals based on the filtered model by one minus the "hat" value from the fit from the original model. This produces a somewhat hybrid computation for R_{pred}^2, but a reasonable number. Of course computation of the other three values, $R_{\mathrm{F}}^2, R_{\mathrm{F,pred}}^2$, and R^2 is unproblematic.

13.6.3   Model Selection After Filtering with a Matrix

13.6.3.1   *Estimation of $\hat{a}$*   The data is monthly, so a saturated periodic model has 12 means. This is facilitated by fitting a periodic function (Fourier series) with six pairs of periodic functions.

The estimate is $\hat{a} = 0.6863$ (Figure 13.13). The assumption is that the saturated model, which has all possible fitting bias removed, will have the most signal-free residuals. This is the model that will be used for estimating $\hat{a}$ for all subsequent models from the periodic family. (Those familiar with *Mallows* C_p might notice a certain similarity of logic.)

13.6.3.2   *The Matrix Representation of a Filter, A*   The modeling process is as follows. The initial unfiltered model is min $\|y - X\beta\|$, where y is a vector of length n, β is a vector of length p, and X is an $n \times p$ matrix. Filtering has the effect of creating a new model: min $\|Ay - AX\beta_{\mathrm{F}}\| \equiv$ min $\|y_{\mathrm{F}} - X_{\mathrm{F}}\beta_{\mathrm{F}}\|$, where the purpose of A, an $(n - 1) \times n$ matrix, is to produce $n - 1$(quasi-)independent observations, y_{F}.

Using a "for" loop to produce filtered values is illuminating, but will quickly become tedious. It usually makes more sense, when p (the columns of X) is large,

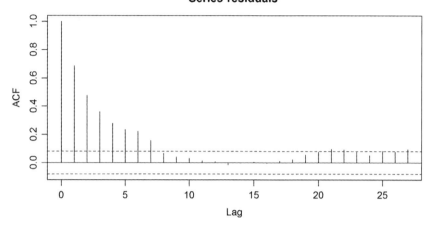

FIGURE 13.13   The *acf( )* function for the residuals from the saturated model.

to construct A and employ matrix multiplication. For an AR(1) model, that matrix A will take the form

$$\begin{bmatrix} 1 & -\hat{a} & 0 & 0 & \dots & 0 \\ 0 & 1 & -\hat{a} & 0 & \dots & \dots \\ 0 & 0 & 1 & -\hat{a} & 0 & \dots \\ \dots & \dots & \dots & \dots & \dots & \dots \\ \dots & \dots & \dots & \dots & 1 & -\hat{a} \end{bmatrix}.$$

Applying the code already developed for Fourier series with variable pairs of trigonometric functions (Section 12.2.2), the following code is added to construct the matrix A and produce a filtered fit to the data.

Suppose a matrix X exists with the number of columns equal to 2·(pairs) + 1. Then the matrix A can be constructed in R as follows:

```
# The A matrix will be almost all zeros. Make a matrix of zeros and add a few changes.
tmp <- rep(0,  n*(n-1))
A <- matrix(tmp, n-1,  n)  # re-shape the vector tmp to a matrix with n-1 rows
                           # and n columns
# in this case, the estimated value of a is 0.6863
a <- 0.6863
for(j in 1:(n-1))
{ A[j,j] <- 1
A[j,j+1] <- -a}
# matrix multiplication to produce the required  model given any
# response vector y and any columns of explanatory variables X
y_f <- A%*%y
X_f <- A%*%X
fit_filtered <- lm(y_f ~ -1 + X_f)
```

TABLE 13.9   The Model Selection Information ($n = 587$) Using Filtered Observations

Model	p	SSE_F	AIC_F	$R^2_{F,pred}(R^2_F)$	$R^2_{pred}(R^2)$
Pairs = 0	1	202.07	3120.16	NA	NA
(single mean)		$(= SST_F)$			
Pairs = 1	3	115.72	2796.92	0.4214	0.6418
				0.4273	0.6454
Pairs = 2	5	44.65	2241.99	0.7752	0.8346
				0.7790	0.8374
Pairs = 3	7	40.07	2182.41	0.7969	0.8402
				0.8017	0.8440
Pairs = 4	9	39.91	2184.06	0.7963	0.7963
				0.8025	0.8442
Pairs = 5	11	39.34	2179.60	0.7978	0.8386
				0.8053	0.8446
Pairs = 6	12	38.72	2172.33[a]	0.7997[a]	0.8383
(saturated)				0.8084	0.8453

[a]Best model.

Any filtered model will have a sample size $n - 1$, a known number of parameters equal to the number of columns of X, and a value for SSE as before. So model selection, on the filtered scale, is quite straightforward.

The saturated model is best (Table 13.8) in this case, and by a wide margin (in terms of AIC differences). On the other hand, in terms of any measure of R^2, any model with at least five parameters seems to work quite well.

Although it might be suggested that the saturated model fits better only because $\hat{a}_1$ has been optimized for this particular model, even if a penalty of two were added to AIC_F, it would still be the best model, so that cannot be the sole source of the better fit of the model.

The last column, R^2_{pred}, reflects how well the model will fit the data on the original scale, but model selection is on the filtered scale using AIC_F or $R^2_{F,pred}$.

13.6.3.3   Model Selection

Model selection information is presented in Table 13.9. There is too much data to plot at once, so only the first 10 years are included in Figure 13.14. Obviously, the more complex models are capturing a lot more of the asymmetric structure (round troughs and pointed peaks, even after logarithmic transformation) in the data. The flexibility of a Fourier series with only a few additional periodic components is quite remarkable.

It is also possible to use likelihood ratio tests. Recall from Chapter 11 that a test for a reduced versus a full model has the form $\chi^2_{p_F-p_R} \sim n \cdot \log_e(SSE_R) - n \cdot \log_e(SSE_F)$.

Table 13.10 contains the pairwise comparisons of various models.

The conclusion from the likelihood ratio tests is essentially the same as that using AIC_F. It is clear that the model with $p = 7$ is superior to all simpler models from the list of candidate models. It is also clear that the saturated model ($p = 12$) is better

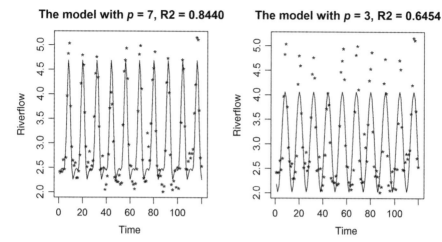

FIGURE 13.14   A good (but not the best) and a poor fit to the data.

than model with four or five pairs of trigonometric components. A final showdown between these two models shows that the saturated model better fits the data.

To reiterate, the model selection exercises in Tables 13.9 and 13.10 are valid because they are performed for (quasi-) independent observations. Without this, or some other correction for serial correlation, these methods cannot be used.

13.7   IMPLICATIONS OF AR(1) ADJUSTMENTS AND THE "SKIP" METHOD

13.7.1   Adjustments for AR(1) Autocorrelation

The value $\sqrt{(1 - \hat{a})/(1 + \hat{a})}$ is not simply the amount by which a test statistic must be reduced, and confidence intervals widened, to account for serial correlation in the data involving means. It is also a number that can be used to develop some rules of thumb and insights about the impact of various levels of autocorrelation.

TABLE 13.10   **The Model Selection Information ($n = 587$) Using Filtered Observations**

Model	p	SSE_F	Likelihood ratio	p-Value
Pairs = 0 (single mean)	1	202.07 (= SST_F)	NA	NA
Pairs = 1	3	115.72	$\chi^2_{3-1} = 329.04$	≈ 0.0
Pairs = 2	5	44.65	$\chi^2_{5-3} = 559.01$	≈ 0.0
Pairs = 3	7	40.07	$\chi^2_{7-5} = 63.53$	≈ 0.0
Pairs = 4	9	39.91	$\chi^2_{9-7} = 2.35$	0.3088
Pairs = 5	11	39.34	$\chi^2_{11-9} = 8.44$	0.0147
Pairs = 6 (saturated)	12	38.72	$\chi^2_{12-11} = 9.32$	0.0095
Supplemental			$\chi^2_{12-7} = 20.12$	≈ 0.0

TABLE 13.11   Reported and True p-Values, When Serial Correlation Is Ignored

$\hat{a}$	Adjustment	$0.10(z = 1.645)$	$0.05(z = 1.960)$	$0.01(z = 2.576)$
0.2	0.816	0.180	0.092	0.036
0.4	0.655	0.281	0.200	0.092
0.6	0.500	0.411	0.327	0.198
0.8	0.333	0.584	0.514	0.391

13.7.2   Impact of Serial Correlation on p-Values

Table 13.11 shows that positive serial correlation, if ignored, p-values thought to be small are actually quite large. For example, if $\hat{a} = 0.4$, a p-value that might naively thought to be 0.05 is actually only 0.20.

13.7.3   The "skip" Method

Of course, filtering is a sophisticated but reliable method for producing (quasi-) independent observations. However, there are times when filtering may not be necessary. In many circumstances, a large number of measurements are taken in time or space. Suppose every other observation, from the original data, was treated as a new set of data. If the autocorrelation between every observation is a^1, and the model for the serial correlation is AR(1), then the autocorrelation between every other observation is a^2. In general, if k observations are skipped, the correlation between the resulting subset of the original data is a^k (an assumption is made that $a > 0$). If a^k can be made sufficiently small, while retaining sufficient sample size, a useable data set of quasi-independent ($a^k < \varepsilon$) observations is produced.

Mathematically, the idea is simple. The goal is to pick a sufficiently large k, so that $\sqrt{(1 + \hat{a}^k)/(1 - \hat{a}^k)} = c$ is close to unity. (The terms a^k and $\hat{a}^k$ are being used somewhat interchangeably here. The theory is based on a^k, but application of the theory, i.e., actual formulas, involve $\hat{a}^k$). Putting these ideas together produces the formula $\log_b \left[(c^2 - 1)/(c^2 + 1) \right] /\log_b(\hat{a}) = k$. This value k is a ballpark estimate of how far apart observations must be to satisfy some notion of quasi independence (the base, b, is irrelevant).

For example, suppose $\hat{a} = 0.45$ and $c = 1.1$ are deemed sufficiently close to unity, then $\log \left[0.21/2.21 \right] /\log(0.45) = 2.9475$, and observations taken three units apart will be sufficiently independent (the number of times to skip, like sample size, should be rounded up to the next integer). In fact, the actual achieved value of c would be $\sqrt{(1 + .45^3)/(1 - .45^3)} = 1.0957 < 1.1$

13.8   SUMMARY

The big picture is that a way of selecting models, in the time series context, is possible for the AR(1) model. A filter is developed and the original data is transformed to produce a related filtered set of data that has the (quasi-) independent property. Analysis of independent data is quite straightforward. The next step is to develop a general strategy for filtering models with errors that are more complex than AR(1). In fact, it is now time to talk about the most general types of error that can result from combinations of AR(m) and MA(l) models.

EXERCISES

1. A conceptual question related to the patch-cut versus uncut watershed case. If there are differences, are they clearly due to patch-cut versus uncut forest? What could be learned by looking at the watersheds associated with several patch-cut and uncut areas. If there is only one patch-cut area, can anything be learned by looking at this area versus several uncut areas?

2. For the simulated data presented in Section 13.3, use the summaries of fit (Table 13.3) to construct a confidence interval for the slope in the filtered and unfiltered model. How much wider is the filtered interval?

3. Justify the comment made in the book, paraphrased as follows: "Because all confidence intervals are too narrow, all p-values are too small."

4. "The lines are not much different." Using the simulated data from Section 13.3, use the summaries of fit (Table 13.3) to make predictions when $x = 5, 25, 45, 65$, and 85 for both the filtered and unfiltered models. Are the predictions similar or different? Explain how these numbers support your conclusion.

5. Produce 2000 simulations like the one given in Section 13.3 and store the estimated values of a. Make a histogram and produce summary statistics for a. How well is a estimated in these simulations.

6. Based on the case in Section 13.3, does the relationship $\sigma^2_{AR(1)}(1 - a^2) = \sigma^2_w$ seem to hold, at least approximately, for this data?

7. A different modeling philosophy attaches more importance to the equal variance assumption when performing t-tests and suggests that the two series in, for example, the logging case have the same underlying AR(1) model so that a better estimate, $\hat{a}$, is $(0.573+0.744)/2 = 0.6585$. Filter the logging data with this common value for $\hat{a}$ and peform a t.test() using the nondefault equal variance assumption (enlist help(t.test) if needed). Is the result different from the book analysis?

8. Describe how the logging case would be carried out as an intervention (Semmelweis) and describe how the Semmelweis case would have been carried out in a fashion similar to the logging case. Further, discuss some threats to inferring cause–effect associated with each design.

9. For the Semmelweis case, analyze the data (producing confidence intervals and a hypothesis test) using the transformation $y = \log_e(p/[1 - p])$. One modification will be required. There is one case where the mortality is zero. A simple adjustment would be to add 1 to each death and 2 to each birth so that

mortality $=$ (deaths $+1$)/(births $+ 2$). There are many sophisticated things that could be done, but this solves the problem while doing little harm to the data. You may choose another adjustment if you feel it is better. Justify the advantage of your method. How does this analysis compare to the example analysis from the book?

10. Create the code for the one- and two-step AR(1) method graphs. Simulate some data with $a = 0.5$ and duplicate all four graphs (Figures 13.11 and 13.12 of Section 13.5)

11. In Section 13.5, it was found that for the one-step method, the width of the interval is $\pm 2\hat{\sigma}_w = \pm 2\sqrt{(1 - \hat{a}^2)}\hat{\sigma}_{AR(1)}$ and, for the two-step method, the width of the interval is $\pm 2\sqrt{(1 + \hat{a}^2)}\hat{\sigma}_w = \pm 2\sqrt{(1 - \hat{a}^4)}\hat{\sigma}_{AR(1)}$. Produce the general formula for the k-step method and, using this formula (it has already been shown in a different manner) show that as $k \to \infty$, the width of the interval becomes $\pm 2\hat{\sigma}_{AR(1)}$.

12. Write the R code to filter the NYC (adjusted) data and produce a summary similar to that given in the book. Compute the correct R^2 for this model.

13. Produce prediction intervals for the next 3 years for the global warming data.

14. If you have had a course involving the definitions of expectation, variance, covariance, and correlation,

   (i) Recall for the unfiltered regression model $y_k = \beta_0 + \beta_1 x_k + \varepsilon_k, \varepsilon_k = a\varepsilon_{k-1} + w_k$ and $\text{Var}(\varepsilon_k) = \sigma^2_{AR(1)}$. Show $\rho^2 = \dfrac{\beta_1^2 \sigma_x^2}{\beta_1^2 \sigma_x^2 + \sigma^2_{AR(1)}}$. R^2 is found by replacing parameters with sample estimates.

   (ii) Recall for the filtered model, $y_k - ay_{k-1} = (1 - a)\beta_0 + \beta_1(x_k - ax_{k-1}) + w_k$, and apply the same technique to get $\rho_F^2 = \dfrac{\beta_1^2 \sigma_x^2}{\beta_1^2 \sigma_x^2 + \frac{1+a}{1-a}\sigma^2_{AR(1)}}$. As before, R_F^2 is found by replacing parameters with estimates.

   (iii) Produce several simulations, with $\rho^2 = 0.80$ and $100 \le n \le 500$ and find R^2 and R_F^2. Do the formulas seem to be working?

15. Using the file "Mitta 36 68.txt" from the Exercises folder (DataMarket–Time Series Data Library–Hydrology–Monthly flows for Mitta Mitta River, Tallandoon, January 1936–December 1968), record monthly flows for the Mitta Mitta river:

   (i) Fit all possible Fourier series to the data and use AIC_F and $R^2_{F,pred}$ to select the best model. Use your own judgment to assess whether a transformation is required.

   (ii) Plot the data with a poor fit and a good fit to the data side by side (use your own judgment to decide how much data to plot, if plotting all of the data produces "too busy" a display).

   (iii) Develop the sequential likelihood ratio tests for this model. Which model appears best using this criterion?

16. Suppose data has been collected and it is known the serial correlation is AR(1) with some fixed a. If samples were taken twice as often, the model would still be AR(1), how would a change? (Challenging). Is there any way the model might not be AR(1) with greater sampling rates?

17. Find the number of observations to skip in the following scenarios. Also, compute the achieved value of c and verify that it meets the criterion:

  (i) $\hat{a} = 0.15, c \leq 1.1$
  (ii) $\hat{a} = 0.25, c \leq 1.1$
  (iii) $\hat{a} = 0.65, c \leq 1.2$
  (iv) $\hat{a} = 0.65, c \leq 1.05$

18. Show $\log_{10}(a)/\log_{10}(b) = \log_e(a)/\log_e(b)$.

PART III

COMPLEX TEMPORAL STRUCTURES

14

THE BACKSHIFT OPERATOR, THE IMPULSE RESPONSE FUNCTION, AND GENERAL ARMA MODELS

14.1 THE GENERAL ARMA MODEL

14.1.1 The Mathematical Formulation

Filtering is the key for fitting many useful models, but now generalization from AR(1) to all kinds of AR(m) and/or MA(l) models is required. The general notion of an ARMA(m,l) model is developed.

The general AR(m) model: $\varepsilon_j = a_1\varepsilon_{j-1} + a_2\varepsilon_{j-2} + \cdots + a_m\varepsilon_{j-m} + j$.

The general MA(l) model: $\varepsilon_j = -b_l w_{j-l} \ldots - b_2 w_{j-2} - b_1 w_{j-2} + w_j$.

The ARMA(m,l) model: $\varepsilon_j = \sum_{s=1}^{m} \alpha_s \varepsilon_{j-s} + \sum_{r=1}^{l} -b_r w_{j-r} + w_j$ where the white noise components have been combined. A rationale will be developed for arguing that, in practice, all such models can be treated as AR(∞), and approximated by AR(m), for some sufficiently large m.

14.1.2 The *arima.sim()* Function in R Revisited

The most general ARMA(m,l) models can be simulated in R using the *arima.sim()* function. For example, consider the ARMA(2,2) model given by $\varepsilon_n = 0.6\varepsilon_{n-1} - 0.25\varepsilon_{n-2} + w_n - 1.1w_{n-1} + 0.28w_{n-2}$. The command to simulate

Basic Data Analysis for Time Series with R, First Edition. DeWayne R. Derryberry.
© 2014 John Wiley & Sons, Inc. Published 2014 by John Wiley & Sons, Inc.

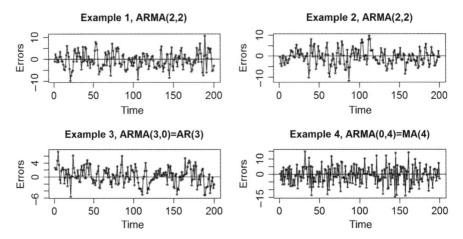

FIGURE 14.1 A representative collection of some ARMA(m,l) model errors.

a series of 200 such errors in R, with a standard deviation of 2.0, is arima.sim(n = 200, list(ar =c(0.6, -0.25),ma = c(1.1, -0.28)), sd = 2.0). Notice that the sign convention in R, for the MA(l) part of the function, is the opposite of this book.

14.1.3 Examples of ARMA(m,l) Models

Four typical error series are displayed below. After developing a bit more theory, it will be possible to develop a taxonomy of the general types of such series. The following series (Figure 14.1) were produced using the following code (the parts with "??" vary from simulation to simulation):

```
error <- arima.sim(n = 200, list(??),sd = 2.0)
plot(1:200, error)
lines(1:200,error)
abline(0,0)
title("An example of ARMA(?,?) errors").
```

Example 1—ARMA(2,2): $\varepsilon_n = 0.6\varepsilon_{n-1} - 0.25\varepsilon_{n-2} + w_n - 1.1w_{n-1} + 0.28w_{n-2}$.
Example 2—ARMA(2,2):$\varepsilon_n = 1.1\varepsilon_{n-1} - 0.28\varepsilon_{n-2} + w_n - 0.6w_{n-1} + 0.25w_{n-2}$.
Example 3—ARMA(3,0):$\varepsilon_n = 0.6\varepsilon_{n-1} - 0.19\varepsilon_{n-2} + 0.084\varepsilon_{n-3} + w_n$.
Example 4—ARMA(0,4): $\varepsilon_n = w_n - 2w_{n-1} + 1.59w_{n-2} - 0.65w_{n-3} + 0.125w_{n-4}$.

The R code (notice that the sign is reversed for $ma()$ compared to the text):

Example 1: arima.sim(n = 200, list(ar=c(0.6, -0.25),ma = c(1.1, -0.28)),sd = 2.0).
Example 2: arima.sim(n = 200, list(ar=c(1.1, -0.28),ma = c(0.6, -0.25)),sd = 2.0).

Example 3: arima.sim(n = 200, list(ar=c(0.6, -0.19,0.084)),sd = 2.0).
Example 4: arima.sim(n = 200, list(ma = c(2.0, -1.59, 0.65, -0.125)),sd = 2.0).

14.2   THE BACKSHIFT (SHIFT, LAG) OPERATOR

14.2.1   Definition of B

The introduction of the backshift operator will allow for integrating all ARMA(m,l) models in a general framework. The shift operator is defined as: $B\varepsilon_n = \varepsilon_{n-1}$, so $B(B\varepsilon_n) = B\varepsilon_{n-1} = \varepsilon_{n-2}$ and, in general, $B^k\varepsilon_n = \varepsilon_{n-k}$. Similarly $B^k w_n = w_{n-k}$.

It will be useful to reexpress some of the basic models in "quasi-polynomial" language using this operator:

AR(1): $(1 - aB)\varepsilon_n = w_n$

AR(2): $\left(1 - a_1 B - a_2 B^2\right)\varepsilon_n = w_n$

AR(m): $a(B)\varepsilon_n = \left(1 - \sum_{j=1}^{m} a_j B^j\right)\varepsilon_n = w_n$ $\left(\text{or } \varepsilon_n = \sum_{j=1}^{m} a_j B^j \varepsilon_n + w_n\right)$.

14.2.2   The Stationary Conditions for a General AR(m) Model

It makes intuitive sense that the stationary condition for the AR(1) model is $|a| < 1$ However, the AR(2) model can now be restated as a combination of two AR(1) models.

Although B is an operator, it is useful to think of polynomials in B.

The "polynomial" associated with the AR(2) model is

$$\left(1 - a_1 B - a_2 B^2\right) = (1 - c_1 B)(1 - c_2 B) = (1 - [c_1 + c_2]B + c_1 c_2 B^2)$$

and $|c_1| < 1$, $|c_2| < 1$ is equivalent to the stability conditions previously stated (the algebra is horrendous).

Similarly, all AR(m) models can be thought of as "polynomials" in B with $a(B) = 1 - \sum_{j=1}^{m} a_j B^j$ (occasionally replacing m with ∞). Polynomials with real coefficients can always be factored $\prod_{j=1}^{m}(1 - c_j B)$, with the usual stipulation that any complex c_j come in conjugate pairs. Stationarity is achieved when $|c_j| < 1$ for all j. Various books on Time Series make reference to a "polynomial in B with all roots outside the unit circle." This is a rather obscure way of saying $|c_j| < 1$ for all j.

For example, $a(B) = 1 - 0.4B + - 0.08B^2 = (1 - [0.2 + 0.2i]B)(1 - [0.2 - 0.2i]B)$ is the "polynomial" associated with a stable AR(2) model. An AR(2) model with complex conjugate values for c_j is called pseudo-periodic (Box et al., 2008, pp. 62–65), with the pseudo-periodic behavior displayed in Figures 5.12 and 8.7.

14.2.3   ARMA(m,l) Models and the Backshift Operator

14.2.3.1   *Representing MA(l) Models with the Backshift Operator*   MA(l) models can also be represented with backshift operators, but on the white noise.

$$\text{MA}(1): \quad \varepsilon_n = (1 - bB)w_n$$

$$\text{MA}(2): \quad \varepsilon_n = (1 - b_1 B - b_2 B^2)w_n$$

$$\text{MA}(l): \quad \varepsilon_n = \left(1 - \sum_{j=1}^{l} b_j B^j\right) w_n = b(B)w_n$$

The most compact representation of an ARMA process is $a(B)\varepsilon_n = b(B)w_n$.

Notice, $a(B)b(B)^{-1}\varepsilon_n = w_n$ suggests that any ARMA(m,l) model can be thought of as AR(∞), provided sense can be made of the expression $b(B)^{-1}$. Although it is not of interest here, it makes just as much or as little sense to argue that $\varepsilon_n = b(B)a(B)^{-1}w_n$ can be used to show all ARMA(m,l) models are MA(∞).

14.2.3.2   *A Diversion—These Are Not Polynomials, But…*   Recall $a(B)$ and $b(B)$ are not really polynomials, because B is an operator. For example, the "polynomial" $a(B)$ does not really have all its roots outside the unit circle, because there are no roots. For example $B = 1/c_j$ is not a root because B is not a variable and does not equal anything. Nevertheless, thinking of $a(B)$ as a polynomial is useful.

For the MA(1) model, the associated polynomial is $b(B) = 1 - c_1 B$. What does $b(B)^{-1}$ mean? Using synthetic division as if it were a polynomial produces the result: $1/(1 - c_1 B) = 1 + c_1 B + c_1^2 B^2 + \cdots = 1 + \sum_{k=1}^{\infty} c_1^k B^k$. This seems like a reasonable result, given that $c_1^k \to 0$, for large k. The suggestion is that MA(1) = AR(∞), and further than MA(1) $\approx$ AR(m) for some large m.

Next, consider the MA(2) polynomial. In this case, $b(B) = 1 - b_1 B - b_2 B^2$ which is factorable to $(1 - c_1 B)(1 - c_2 B)$. When the roots are real, using partial fractions, it is easy to produce the resulting series.

With unequal roots, $1/(1 - c_1 B)(1 - c_2 B) = \frac{-c_1/(c_2 - c_1)}{1 - c_1 B} + \frac{c_2/(c_2 - c_1)}{1 - c_2 B}$, and this, from the previous analysis, is the sum of two convergent infinite series.

With equal roots $1/(1 - c_1 B)^2 = (1 + c_1 B + c_1^2 B^2 + \cdots)^2$. In other words, it appears MA(2) = AR(∞) in this case as well (the case of complex roots is quite messy).

It is now possible to think in more general terms about construction of ARMA(m,l) errors. Series can be characterized by whether the underlying roots are real or complex conjugates and whether the real roots are negative or positive (a fact alluded to in Chapters 5 and 6).

14.2.4   More Examples of ARMA(m,l) Models

The examples from Section 14.1 can be characterized as follows:

Example 1: The AR(2) model had complex roots ($0.3 \pm 0.4i$), while the MA(2) model had real positive roots (0.4, 0.7).

Example 2: The roots were reversed. The MA(2) model had complex roots and the AR(2) model had positive, real roots.

Example 3: The AR(3) model had positive, real roots $(0.3, 0.4, 0.7)$.

Example 4: The MA(4) model had two pairs of complex roots $(0.3 \pm 0.4i, 0.7 \pm 0.1i)$.

A few more examples are given, while others are assigned as exercises. All simulations used arima.sim(n = 100,...,sd = 1.0).
These examples appear in Figure 14.2.

Example 5: AR(2) with roots $(0.8, 0.1)$ and MA$(0.25, 0.75)$:

$$(1 - 0.9B + 0.08B^2)\varepsilon_n = (1 - B + 0.1875B^2)w_n.$$

Example 6: AR(2) with roots $(-0.6, 0.2)$ and MA(2) with roots $(0.1 \pm 0.8i)$:

$$(1 + 0.4B + 0.12B^2)\varepsilon_n = (1 - 0.2B + 0.65B^2)w_n.$$

Example 7: AR(2) with roots $(-0.4, 0.8)$ and MA(2) with roots $(0.75, 0.75)$:

$$(1 - 0.4B - 0.32B^2)\varepsilon_n = (1 - 1.5B + 0.5625B^2)w_n.$$

Example 8: AR(4) with roots $(0.3 \pm 0.4i, -0.7 \pm 0.1i)$:

$$(1 + 0.8B + 0.09B^2 + 0.05B^3 + 0.125B^4)\varepsilon_n = (1)w_n.$$

The variety of potential behavior of the errors is quite diverse and apparently quite unpredictable. It is not obvious to the author why examples 5 and 7 are alike

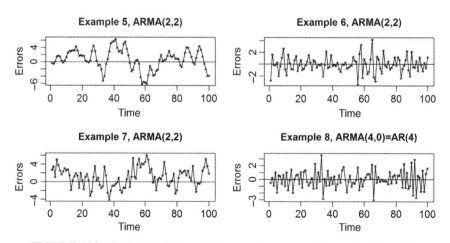

FIGURE 14.2   Four more representative errors from general ARMA(m,l) models.

and examples 6 and 8 are alike, but the odd and even examples are so strikingly different. The similarities and differences among these examples were not planned by the author.

14.3   THE IMPULSE RESPONSE OPERATOR – INTUITION

An interesting question is about what happens when a system in stable equilibrium receives a shock (impulse). How does it respond to return to equilibrium?

Consider a general AR(2) model and the following scenario: For some fixed k, for $j \neq k$, $\varepsilon_j = a_1\varepsilon_{j-1} + a_2\varepsilon_{j-2} + w_j$; and for $j = k$, $\varepsilon_j = a_1\varepsilon_{j-1} + a_2\varepsilon_{j-2} + w_j + c$. The shock, c, will be dampened out of the errors in a predictable manner.

Some tedious algebra, using iterative substitutions, produces the exact pattern. In what follows, it is important to recall that for all $j < k$, the shock has not yet occurred and that later white noise components are also unaffected by the shock:

$$\varepsilon_k = a_1\varepsilon_{k-1} + a_2\varepsilon_{k-2} + w_k + c$$

$$\varepsilon_{k+1} = a_1^2\varepsilon_{k-1} + (a_1a_2 + a_2)\varepsilon_{k-2} + a_1w_k + w_{k+1} + a_1c$$

$$\varepsilon_{k+2} = \left(a_1^3 + a_1a_2\right)\varepsilon_{k-1} + \left(a_1^2a_2 + a_1a_2 + a_2^2\right)\varepsilon_{k-2} + \left(a_1^2 + a_2\right)w_k$$
$$+ a_1w_{k+1} + w_{k+2} + \left(a_1^2 + a_2\right)c$$

$$\varepsilon_{k+3} = \left(a_1^4 + a_1^2a_2 + a_1a_2^2\right)\varepsilon_{k-1} + \left(a_1^3a_2 + a_1^2a_2 + 2a_1a_2^2 + a_2^2\right)\varepsilon_{k-2}$$
$$+ \left(a_1^3 + 2a_1a_2\right)w_k + \left(a_1^2 + a_2\right)w_{k+1} + a_1w_{k+2} + w_{k+2} + \left(a_1^3 + 2a_1a_2\right)c.$$

A recursive formula, which will be introduced shortly, can be used to compute all of the coefficients above. Because $|a_j| < 1$, the shock disappears from the sequence over time—this is another aspect of stable equilibrium (stationarity).

Let $a_1 = 0.4$ and $a_2 = 0.21$, then a simulation displays the behavior of the error and the shock:

```
# produce 20 observations with no shock
pre_error <- arima.sim(n = 20, list(ar=c(0.4, 0.21)), sd = 2.0)
# introduce a large, one-time shock at time period 21
shock <- 25 + 0.4*pre_error[20]+0.21*pre_error[19] + rnorm(1,0,2)
# return to the AR(2) model
error <- rep(NA, 200)
error[1:20] <- pre_error
error[21] <- shock
for(j in 22:200)
{ error[j] <- 0.4*error[j-1] + 0.21*error[j-2] + rnorm(1,0,2)}
```

The manner in which the dampening occurs (Figure 14.3) is related to the impulse response function.

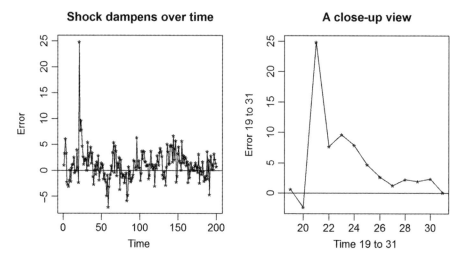

FIGURE 14.3   The series (left) and a close-up look at observations 19:31 (right).

14.4   IMPULSE RESPONSE OPERATOR, *g(B)*—COMPUTATION

14.4.1   Definition of *g(B)*

The impulse response operator (sometimes the term impulse response function is used, but what is first derived is not a function) is a way of writing the sequence of errors as a function of all past white noise elements. For all ARMA(m,l) models:
$\varepsilon_n = \frac{b(B)}{a(B)} w_n = g(B) w_n = \sum_{j=0}^{\infty} g_j w_{n-j}$.

14.4.2   Computing the Coefficients, g_j

For a concrete application, the values are computed directly from the definition, by matching coefficients on the different "powers" of B^k. In other words,

$a(B) g(B) = b(B)$, so ...
$\left(1 - a_1 B - a_2 B^2 + ...\right)\left(g_0 + g_1 B + g_2 B^2 + ...\right) = \left(1 - b_1 B - b_2 B^2 - ...\right).$

A recursive relationship produces $g_0 = 1$ and $g_k = \sum_{j=1}^{k} a_j g_{k-j} - b_k$ for $k = 1, 2, 3 ...$

For example, $g_3 = a_1 g_2 + a_2 g_1 + a_3 g_0 - b_3$.
Some simple cases

AR(1) models: In this case, $g_0 = 1$ and $g_k = a g_{k-1} = a^k$ for all $k = 1, 2, 3 ...$ In this case, the impulse response function shows an exponential decay ($a > 0$) or an oscillating decay ($a < 0$).

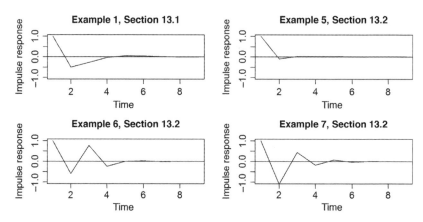

FIGURE 14.4   Typical impulse response functions for ARMA(2,2) models.

ARMA(2,2) models: In this case, $g_0 = 1$, $g_1 = a_1 - b_1$, $g_2 = a_1 g_1 + a_2 - b_2$

$$g_k = a_1 g_{k-1} + a_2 g_{k-2} \text{ for } k = 3, 4, 5 \dots$$

AR(2) models: These are just ARMA(2,2) models with $b_1 = b_2 = 0$.

14.4.3   Plotting an Impulse Response Function

The impulse response operator can be used to express any error as a function of all past white noise elements, the impulse response function. It is straightforward to write a program implementing the recursive relationships, in general, and taking advantage of special cases.

For example, a general program for an ARMA(2,2) model could be coded as follows:

```
a1 <- 0.4
a2 <- 0.32
b1 <- 1.5
b2 <- -0.5625
g <- rep(NA,19)
g[1] <- a1-b1
g[2] <- a1*g[1]+a2 -b2
for(m in 3:19)
  { g[m] <- a1*g[m-1] + a2*g[m-2] }
g <- c(1,g) # add g0 at the end
plot(1:20,g)
lines(1:20,g)
abline(0,0)
title("The impulse response function for Example 7, Section 14.2")
```

This code produces the plots in Figure 14.4.

The impulse response function represents the dampening of the shock at time $j+1$, c, as follows:

$$\varepsilon_j = g(B)w_j = g_0 w_j + g_1 w_{j-1} + g_2 w_{j-2} + g_3 w_{j-3} + \cdots$$

$$\varepsilon_{j+1} = g(B)(w_{j+1} + c) = g_0 w_{j+1} + g_1 w_j + g_2 w_{j-1} + g_3 w_{j-2} + \cdots + g_0 c$$

$$\varepsilon_{j+2} = g(B)w_{j+2} = g_0 w_{j+2} + g_1 w_{j+1} + g_2 w_j + g_3 w_{j-1} + \cdots + g_1 c \text{ , etc.}$$

In each following time period, the series with the shock becomes $\varepsilon_{j+r} = g(B)w_{j+r} + g_r c$.

As long as $g_r \rightarrow 0$ for large r, the shock will dampen to zero.

14.5   INTERPRETATION AND UTILITY OF THE IMPULSE RESPONSE FUNCTION

The impulse response function will be used to derive the covariance structure and the spectrum for all ARMA(m,l) models. These results will allow the derivation of the Yule–Walker equations and the last plot (and R function) of interest—the partial autocorrelation plot [using *pacf()* in R].

The impulse response function also has a useful interpretation. Because $\varepsilon_n = g(B)w_n$, every ARMA(m,l) sequence or error can be thought of as a linear combination of all past white noise, with g_k, the weight attached to the white noise component w_{n-k}. It should be intuitively obvious that any series that is useful in the real world must have $g_k \rightarrow 0$, for large k. That is, the current error should be dependent on recent events, not events in the ancient past. Although not proven here, the graphs show that stationary and invertible ARMA(2,2) errors have impulse response functions with this characteristic.

EXERCISES

1. Find the impulse response function for (i) MA(1), (ii) MA(2), (iii) white noise, and (iv) AR(3)

2. For the model ARMA(3,0) with c_j $0.7 \pm 0.1i$ and 0.65:
   (i) Find $a(B)$ and $b(B)$, (ii) simulate and plot the data, (iii) how is it known the model is stationary?

3. For the model ARMA(4,0) with c_j $0.7 \pm 0.1i$, 0.45, and 0.65:
   (i) Find $a(B)$ and $b(B)$, (ii) simulate and plot the data, (iii) how is it known the model is stationary?

4. For the model ARMA(0,4) with c_j $0.7 \pm 0.1i$, -0.45, and 0.65:
   (i) Find $a(B)$ and $b(B)$, (ii) simulate and plot the data, (iii) how is it known the model is invertible?

5. For the model ARMA(2,2) with AR(2) c_j $0.7 \pm 0.1i$, and MA(2) c_j -0.45, and 0.65:
   (i) Find $a(B)$ and $b(B)$, (ii) simulate and plot the data, (iii) how is it known the model is invertible and stationary?

6. For the model ARMA(2,2) with AR(2) c_j $0.7 \pm 0.1i$, and MA(2) c_j $0.2 \pm 0.6i$:
   (i) Find $a(B)$ and $b(B)$, (ii) simulate and plot the data, (iii) how is it known the model is invertible and stationary?

7. For the model ARMA(2,2) with AR(2) c_j 0.7 and -0.6, and MA(2) c_j 0.2, and -0.8:
   (i) Find $a(B)$ and $b(B)$, (ii) simulate and plot the data, (iii) how is it known the model is invertible and stationary?

8. Find the impulse response function for the following and write a program to plot the first 20 values (g_0 to g_{19}) :
   (i) Exercise 2, (ii) Exercise 3, (iii) Exercise 4, (iv) Exercise 5, (v) Exercise 6

9. Consider an AR(2) model and the required stability conditions.
   (i) Find values of a_1 and a_2 that violate one and only one of the three stationarity conditions (you should have three sets of values, one for each violation).
   (ii) Plot the first 25 values for the impulse response functions.
   (iii) What do you conclude?

15

THE YULE–WALKER EQUATIONS AND THE PARTIAL AUTOCORRELATION FUNCTION

15.1 BACKGROUND

In this chapter, the last tools are developed in order to justify modeling any ARMA(m,l) model with an AR(m) filter: The Yule–Walker equations [and the R functions *ar.yw( )* and *ar.mle( )*], the partial autocorrelation function and plot [and the R function *pacf( )*], and the spectrum for ARMA(m,l) models as well as the relationship between the spectrum and the impulse response function (this will not be particularly useful for modeling, but it is too cool not to mention, and does not take up much space).

15.2 AUTOCOVARIANCE OF AN ARMA(m,l) MODEL

15.2.1 A Preliminary Result

The autocovariance function, with the derivation of the impulse response function, can be viewed from a very different perspective. Recall that there is an interest in computing things like $E(w_r \varepsilon_t)$, but now it has been established that ε_t is a linear combination of all white noise components up to, and including, time t. Recall also that $E(w_j w_k) = 0$, unless $j = k$, in which case $E(w_j^2) = \sigma_w^2$.

Basic Data Analysis for Time Series with R, First Edition. DeWayne R. Derryberry.
© 2014 John Wiley & Sons, Inc. Published 2014 by John Wiley & Sons, Inc.

So $w_r \varepsilon_t = w_r g(B) w_t = \sum_{j=0}^{\infty} g_j w_r w_{t-j}$ and $E(w_r \varepsilon_t) = \sum_{j=0}^{\infty} g_j E(w_r w_{t-j})$. The only nonzero component of this infinite series is when $r = t - j$, if that is possible (if $r > t$, this is not possible). Therefore:

$$E(w_r \varepsilon_t) = \begin{cases} 0, & r > t \\ g_{t-r} \sigma_w^2, & r \leq t. \end{cases}$$

15.2.2   The Autocovariance Function for ARMA(*m,l*) Models

Recall that the autocovariance, $C_k = C_{-k} = E(\varepsilon_j, \varepsilon_{j \pm k})$. A few definitions and the application of the previous results yields

$$\varepsilon_j = \sum_{s=1}^{m} a_s \varepsilon_{j-s} - \sum_{p=1}^{l} b_p w_{j-p} + w_j$$

$$\varepsilon_j \varepsilon_{j-k} = \sum_{s=1}^{m} a_s \varepsilon_{j-s} \varepsilon_{j-k} - \sum_{p=1}^{l} b_p w_{j-p} \varepsilon_{j-k} + w_j \varepsilon_{j-k}$$

$$C_k = E(\varepsilon_j \varepsilon_{j-k}) = \sum_{s=1}^{m} a_s E(\varepsilon_{j-s} \varepsilon_{j-k}) - \sum_{p=1}^{l} b_p E(w_{j-p} \varepsilon_{j-k}) + E(w_j \varepsilon_{j-k})$$

Suppose $k = 0$,

$$C_0 = \sum_{s=1}^{m} a_s E(\varepsilon_{j-s} \varepsilon_j) - \sum_{p=1}^{l} b_p E(w_{j-p} \varepsilon_j) + E(w_j \varepsilon_j)$$

$$= \sum_{s=1}^{m} a_s C_s - \sum_{p=1}^{l} b_p g_p \sigma_w^2 + \sigma_w^2.$$

Suppose $k > 0$

$$C_k = \sum_{s=1}^{m} a_s C_{k-s} - \sum_{p=1}^{l} b_p g_{p-k} \sigma_w^2 + 0.$$

In the second summation $g_{p-k} = 0$ when $k > p$.

15.3   AR(*m*) AND THE YULE–WALKER EQUATIONS

15.3.1   The Equations

For any AR(*m*) model, $b_p = 0$, for all p. In this special case, the previous equations reduce to the Yule–Walker equations (replacing parameters with estimates):

$$\hat{C}_{0,\text{AR}} = \sum_{j=1}^{m} \hat{a}_j \hat{C}_j + \hat{\sigma}_w^2 \text{ and } \hat{C}_{k,\text{AR}} = \sum_{j=1}^{m} \hat{a}_j \hat{C}_{k-j}.$$

In other words, if the autocovariance function has been computed, and it is assumed the model is ARMA(m,0), it is possible to estimate the unknown values, $\hat{a}_1 \dots \hat{a}_m, \hat{\sigma}_w^2$ by solving a linear system of equations.

The AR(1) model, using the Yule–Walker equations, produces the system $\hat{C}_0 = \hat{a}_1\hat{C}_1 + \hat{\sigma}_w^2$ and $\hat{C}_1 = \hat{a}_1\hat{C}_0$. From the *acf()* function, the values $\hat{C}_0$ and $\hat{C}_1$ can be computed, which will allow estimation of a_1 and σ_w^2 using the Yule–Walker equations. In matrix form, the system is

$$\begin{pmatrix} \hat{C}_0 \\ \hat{C}_1 \end{pmatrix} = \begin{pmatrix} \hat{C}_1 & 1 \\ \hat{C}_0 & 0 \end{pmatrix} \begin{pmatrix} \hat{a}_1 \\ \hat{\sigma}_w^2 \end{pmatrix}$$

Similarly, for an AR(2) model, $m = 2$, there are three equations in three unknowns:

$$\hat{C}_0 = \hat{a}_1\hat{C}_1 + \hat{a}_2\hat{C}_2 + \hat{\sigma}_w^2, \ \hat{C}_1 = \hat{a}_1\hat{C}_0 + \hat{a}_2\hat{C}_{-1}, \text{ and } \hat{C}_2 = \hat{a}_1\hat{C}_{-1} + \hat{a}_2\hat{C}_{-2}.$$

In matrix form [recall that $C_k = C_{-k}$ and *acf()* is used to estimate values], the system is

$$\begin{pmatrix} \hat{C}_0 \\ \hat{C}_1 \\ \hat{C}_2 \end{pmatrix} = \begin{pmatrix} \hat{C}_1 & \hat{C}_2 & 1 \\ \hat{C}_0 & \hat{C}_1 & 0 \\ \hat{C}_1 & \hat{C}_2 & 0 \end{pmatrix} \begin{pmatrix} \hat{a}_1 \\ \hat{a}_2 \\ \hat{\sigma}_w^2 \end{pmatrix}.$$

15.3.2 The R Function *ar.yw()* with an AR(3) Example

An R function is available for solving the Yule–Walker equations, *ar.yw()*. Pass any sequence of errors or residuals to the function and the function will solve a sequence of models, AR(1)…AR(m) and pick the best model using the criterion of minimizing AIC. It should be noted that the Yule–Walker estimates are not maximum likelihood estimates, so that AIC is only approximated.

This useful R function will be demonstrated by first simulating a sequence of AR(3) errors and comparing the fitted models AR(0),…, AR(6). Subtle distinctions are important. Because the true model is known to be AR(m), BIC should be used to select the best model. Hypothesis testing will also be used as an alternative method of model selection.

The simplest way to be assured that the AR(3) model is stationary is to begin with the roots, being assured, they are all less than one in magnitude.

$$(1 - 0.7B)(1 + 0.2B)(1 - 0.9B)\varepsilon_n = (1 - 1.4B + 0.31B^2 + 0.126B^3)\varepsilon_n = w_n$$

```
# Simulating 2000 AR(3) errors
error <- arima.sim(n = 2000, list(ar=c(1.4,-0.31, -0.126)),sd = 3.0)
plot(1:2000,error)
lines(1:2000, error)
abline(0,0)
title("AR(3) errors, n= 2000")
```

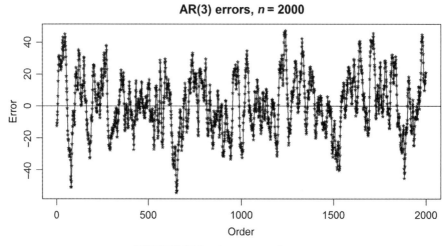

FIGURE 15.1 A sequence of errors.

This produces Figure 15.1.

The next step is to use the function *ar.yw( )* and understand the various information produced by the function:

```
z <- ar.yw(error), names(z)
```

[1] "order"	"ar"	"var.pred"	"x.mean"	"aic"
[6] "n.used"	"order.max"	"partialacf"	"resid"	"method"
[11] "series"	"frequency"	"call"	"asy.var.coef"	

The most important command, following up on ar.yw() is z$aic

0	1	2	3	4	5
7218.086359	632.339343	33.650225	0.000000	0.941779	2.716482

6	7	8	9	10	11
4.697557	6.140760	6.326040	8.076309	9.674298	11.588804...

This is a common format for AIC values. With AIC (or BIC), the values themselves are unique only up to a common constant, and it is the minimum value and differences that are important. Therefore, all values are frequently reported as differences from the minimum value. Using AIC, the AR(3) model is identified as the best, and the AR(4) model is only slightly worse (0.941779 greater). In some sense, the AR(3) model appears just 1.6 (exp[0.941779/2] = 1.6) times more believable than the AR(4) model, not a large difference.

Some other components of this output often used are

The order, m, of the model chosen by AIC: z$order 3.

The values of $\hat{a}_1 \ldots \hat{a}_m$ for the chosen order: z$ar 1.4133773 -0.3124876 -0.1329180.

The estimate $\hat{\sigma}_w^2$ (recall, $\sigma_w^2 = 9$): z$var.pred 8.728231.

In some cases, there is a desire to find $\hat{a}_1 \ldots \hat{a}_m$ and/ or $\hat{\sigma}_w^2$ for a different model. In that case, the command z <- ar.yw(error, aic = FALSE, order.max = m) will ignore AIC (aic = FALSE) and force the fitting of some specified order m. [Recall, many other things can be learned using help(ar.yw)].

15.3.3   Information Criteria-Based Model Selection Using *ar.yw()*

For the previous simulation, the correct model is to be picked from a list of candidate models. Therefore, in this circumstance, BIC is preferred to AIC for model selection (Aho et al., 2014). As with many previous modeling exercises, a model selection table is developed.

For each candidate model, AR(0) to AR(7), the command ar.yw(error, aic = FALSE, order.max = m) can be used to fit a specified order [For $m = 0$, var(error) was used to estimate $\hat{\sigma}_w^2$]. Because information criteria assumes maximum likelihood estimates, a slightly different command can be used to get just those values—z <- ar.mle(error, aic = FALSE, order.max = m). The model selection table is produced using *ar.mle()* to obtain $\hat{\sigma}_w^2$.

It is not obvious precisely how AIC is computed for *ar.yw()*, nor is BIC computed. However, a reasonable estimate would be BIC $\approx n \cdot \log_e(\hat{\sigma}_w^2) + m \cdot \log_e(n)$, using *ar.mle()* to estimate σ_w^2. The estimated BIC values in Table 15.1 were computed in this manner.

The choice of *ar.yw()* versus *ar.mle()* for estimation of σ_w^2 was of no impact [The AIC values here match the earlier *ar.yw()*-based values almost perfectly]. Because *ar.mle()* may run slow or even fail to converge for large sets of data, and

TABLE 15.1   Model Selection Using Information Criteria ($n = 2000$)

Model	$\hat{\alpha}_w^2$	ΔAIC	$\approx$ BIC	$\approx \Delta$BIC
$m = 0$	322.82	7218.09	11,554.19	7202.44
1	11.97	632.34	4972.41	620.66
2	8.87	33.65	4380.55	28.80
3	8.71	0.00	4351.75	0.00
4	8.70	0.94	4357.05	5.30
5	8.70	2.72	4364.65	12.90
6	8.70	4.70	4372.25	20.50
7	8.70	6.14	4379.85	28.80

because the derivation for the Yule–Walker equations has been given and the derivation of the maximum likelihood approach is beyond this book, $ar.yw()$ will be the default method for the rest of the book.

The results from Table 15.1 correctly identify the AR(3) model as the model that generated that data. Of course, in this case, the choice is obvious and both AIC and BIC make the same correct choice. The estimated variance drops substantially for each order up to and including order 3. However, after that there is no further drop in variance, so further increases in complexity gain nothing.

15.4 THE PARTIAL AUTOCORRELATION PLOT

15.4.1 A Sequence of Hypothesis Tests

Let $\hat{a}_m^m$ be the estimate of $\hat{a}_m$ for the model AR(m). It is known that (Box et al., 2008, p. 70) $\hat{a}_m^m \sim N(0, 1/n)$ if the true model has order less than m. Therefore $z \approx \sqrt{n} \cdot \hat{a}_m^m$ is standard normal if the order is less than m and can be treated as a test statistic for "Ho: the order is less than m" versus "Ha: the order is at least m."

Table 15.2 continues with the last data, adding this computed z-score and the p-value associated with the z-score.

It seems clear that the correct order is 3 from these hypothesis tests. Put more precisely, the data suggests that an order of at least 3 is required, but there is no evidence that an order higher than 3 is required. In this case, AIC, BIC, and hypothesis tests produce identical and unambiguous model selection. Table 15.2 also shows how similar the estimates provided by $ar.yw()$ and $ar.mle()$ often are, at least for large data sets.

15.4.2 The $pacf()$ Function—Hypothesis Tests Presented in a Plot

If the values $\hat{a}_m^m$ for $m = 1,2,3,\dots$ are plotted versus m, with lines at $\pm 2/\sqrt{n}$, the display becomes a visual representation of the above hypothesis tests. When $\hat{a}_m^m$ falls far outside the lines $\pm 2/\sqrt{n}$, there is evidence that the model is at least of order m; when the values begin to fall far inside the lines, the indications are that the order is no higher than the last large spike.

TABLE 15.2 A Sequence of Tests for the Order ($n = 2000$)

AR(m)	$\hat{a}_m^m$ (yw)	$\hat{a}_m^m$ (mle)	z	p-Value
$m = 1$	0.9813	0.9808	43.86	$\approx$ 0.0000
2	−0.5093	−0.5089	−22.76	$\approx$ 0.0000
3	−0.1329	−0.1335	−5.97	$\approx$ 0.0000
4	0.0223	0.0225	1.01	0.3124
5	0.0106	0.0109	0.49	0.6242
6	0.0031	0.0032	0.14	0.8886
7	−0.0167	−0.0161	0.72	0.4716

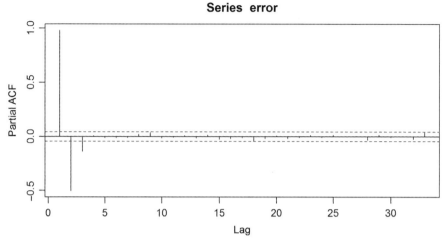

FIGURE 15.2 The *pacf( )* plot for the previously simulated errors.

It is clear from Figure 15.2, which is the information from Table 15.2 in a slightly different format, that the errors appear to be of order $m = 3$.

The autocorrelation plot and the partial autocorrelation plot have complimentary roles. For an AR(m) process, the partial autocorrelation plot is expected to have m significant spikes while the spikes in the autocorrelation plot will slowly decay. For an MA(l) process, the autocorrelation plot will have l distinct spikes, but the partial autocorrelation plots will exhibit a low decay (Box et al., 2008, p. 75; Kitagawa, 2010, pp. 87–89). Compare the plots for the following errors:

```
AR_error  <- arima.sim(n = 500, list(ar=c(0.5,0.4)), sd = 2.0)
MA_error <- arima.sim(n = 500, list(ma=c(0.5,0.4)), sd = 2.0)
```

The *pacf( )* and *acf( )* plots for these errors are displayed in Figure 15.3.

As can be seen from Figure 15.3, the theory often finds the data quite uncooperative.

15.5 THE SPECTRUM FOR ARMA PROCESSES

Recall that the spectrum is defined as

$$p(f) = \sum_{k=-\infty}^{\infty} C_k \cdot \exp(-2\pi k f i)$$

The expansion $C_k = E(\varepsilon_j \varepsilon_{j-k}) = E\left\{ \sum_{s=0}^{\infty} g_s w_{j-s} \cdot \sum_{p=0}^{\infty} g_p w_{j-k-p} \right\}$ is now available using the impulse response function.

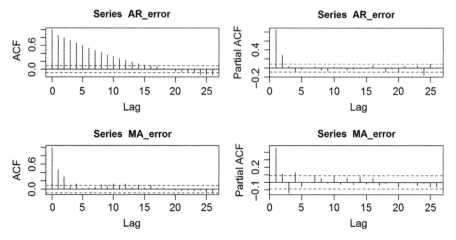

FIGURE 15.3   MA(l) versus AR(m), *acf( )* and *pacf( )* plots.

Note, as before $g_p = 0$, for $p < 0$. After some preliminary manipulations, C_k will be found to be a function of the g_j values, and these will be substituted into the above expression for $p(f)$. A surprising and elegant result will follow:

$$E\left\{ \sum_{s=0}^{\infty} g_s w_{j-s} \cdot \sum_{p=0}^{\infty} g_p w_{j-k-p} \right\} = \sum_{s=0}^{\infty} \sum_{p=0}^{\infty} g_s g_p E(w_{j-s} \cdot w_{j-k-p}).$$

Recall that $E(w_r w_t) = \sigma_w^2$, when $r = t$, and $E(w_r w_t) = 0$, when $r \neq t$. Therefore, all terms in the above expression are zero unless $p = k - s$. So

$$C_k = \sigma_w^2 \sum_{s=0}^{\infty} \sum_{p=0}^{\infty} g_s g_p = \sigma_w^2 \sum_{s=0}^{\infty} g_s g_{k-s} \text{ , recall that } g_{k-s} = 0, \text{ when } k < s.$$

Therefore $p(f) = \sum_{k=-\infty}^{\infty} C_k \exp(-2\pi k f i) = \sigma_w^2 \sum_{k=-\infty}^{\infty} \sum_{s=0}^{\infty} g_s g_{k-s} \exp(-2\pi k f i).$

$$\sigma_w^2 \sum_{s=0}^{\infty} \sum_{k=-\infty}^{s} g_s g_{k-s} \exp(-2\pi k f i) = \sigma_w^2 \sum_{s=0}^{\infty} \sum_{k=-\infty}^{s} g_s \exp(-2\pi s f i) g_{k-s} \exp(-2\pi [k - s] i).$$

Recall that $g_{k-s} = 0$ for $k - s < 0$.
Now let $p = k - s$ (Figure 15.4).

$$\sigma_w^2 \sum_{s=0}^{\infty} g_s \exp(-2\pi s f i) \cdot \sum_{p=0}^{\infty} g_p \exp(2\pi p f i) = \sigma_w^2 \left| \sum_{p=0}^{\infty} g_p \exp(-2\pi p f i) \right|^2.$$

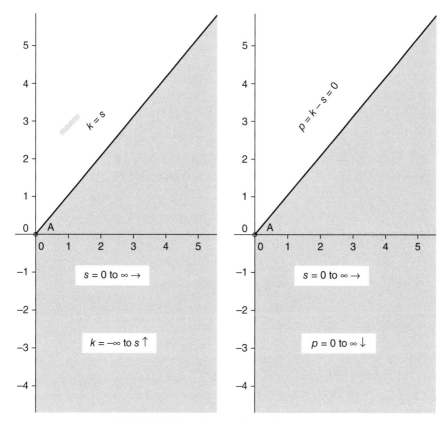

FIGURE 15.4   Change of variable, $p = k - s$.

If this were not interesting enough in itself, recall that $g(B) = a(B)^{-1}b(B)$. Furthermore, the method of equating like coefficients to derive all g_j for $j = 0,1,2,\ldots$ is valid when B is replaced with $\exp(-2\pi tfi)$. So

$$\sum_{t=0}^{\infty} g_t e^{-2\pi fi} = \left[1 - \sum_{t=1}^{l} b_t e^{-2\pi tfi}\right] \bigg/ \left[1 - \sum_{t=1}^{m} a_t e^{-2\pi tfi}\right] \quad \text{and}$$

$$p(f) = \sigma_w^2 \left|\sum_{t=0}^{\infty} g_t e^{-2\pi tfi}\right|^2 = \sigma_w^2 \frac{\left|1 - \sum_{t=1}^{l} b_t e^{-2\pi tfi}\right|^2}{\left|1 - \sum_{t=1}^{m} a_t e^{-2\pi tfi}\right|^2}.$$

15.6   SUMMARY

Chapter 14 introduced a substantial generalization and integration of the ARMA(m,l) family of nonwhite noise error. In Chapter 14, the impulse response function was developed, as was an informal argument that ARMA(m,l) = AR(∞). In Chapter 15, these ideas were extended so that it was shown that a broad number of tools are possible for examining AR(m) models. It has been shown that, for all AR(m) models, the Yule–Walker equations implemented in R with the function *ar.yw()* allow for estimation of all the values $a_1 \ldots a_m$ and σ_w^2, with model selection based on AIC. Also, the R function *pacf()* provides model selection based on a sequence of hypothesis tests presented in a graph that has the property that, for any AR(m) model, the graph will have exactly m spikes. It is also hinted, in Chapter 13, that a filter can be developed for any AR(m) model.

Given the wealth of tools for AR(m) models and the lack of tools for MA(l) models, it may not be hard to guess where the next chapter leads.

EXERCISES

1. Show that, for complex numbers, if $f = g/h$, then $\bar{f} = \bar{g}/\bar{h}$ (a brute force approach involves defining g and h quite generally and showing the result).

2. Using *pacf()* and *ar.yw()* or *ar.mle()*, verify that the following models from Chapter 13 are, indeed, quite plausibly AR(1). That is, the residuals are AR(1) after fitting the signal: (i) the logging data, (ii) the global warming data, (iii) the Semmelweis data, (iv) the adjusted NYC temperatures, (iv) Boise riverflow data (with at least three pairs of periodic functions).

3. Simulate 400 AR(1) errors with $a = 0.8$ and $\sigma = 2.0$. Use *ar.yw()* to get the estimates of a and σ and verify that these estimates satisfy the Yule–Walker equations.

4. Repeat Exercise 3 for an AR(2) model with $a_1 = 0.7$ and $a_2 = -0.3$.

5. Simulate MA(2) data with $b_1 = -0.5$, $b_2 = 0.1$, and $n = 500$. Find the autocorrelation plot and partial autocorrelation plot. Do the plots follow the pattern suggested in this chapter? Explain.

6. Simulate AR(2) data with $a_1 = 0.7$, $a_2 = -0.4$, and $n = 400$. Find the autocorrelation plot and partial autocorrelation plot. Do the plots follow the pattern suggested in this chapter? Explain.

7. Give the matrix representation for the Yule–Walker equations for AR(4).

8. Using $(1 - .7B)(1 + [0.1 - 0.1i]B)(1 - [0.1 + 0.1i]B)\varepsilon_n = w_n$
   - **(i)** Find the values for a_1, a_2, and a_3.
   - **(ii)** Why do we know this is a stationary process?
   - **(iii)** Write out the Yule–Walker equations for AR(3).
   - **(iv)** Simulate $n = 500$ observations from this model.
   - **(v)** Using *ar.yw()* [or *ar.mle()*] and *pacf()*, assess whether the data appears to be AR(3).

(vi) Using ar.yw(x, aic = FALSE, order.max = 3) verify that the values computed with this routine satisfy the Yule–Walker equations (expect an unavoidable round-off error on your part).

9. Consider the following three AR(2) (quadratic) problems of the form

$$a(B) = (1 - c_1B)(1 - c_2B).$$

Case 1: $c_1 = 0.7, c_2 = 0.6$
Case 2: $c_1 = 0.7, c_2 = -0.6$
Case 3: $c_1 = 0.6 - 0.3i, c_2 = 0.6 + 0.3i$

in each case, complete the following steps:

(i) Determine the coefficients a_1 and a_2.
(ii) Simulate the data according to this model, with $n = 800$.
(iii) Fit models A(1)...AR(4) to the data using the *ar.yw()* or *ar.mle()*.
(iv) Complete a table that uses AIC, BIC, and hypothesis testing to select the best model.
(v) Verify, using *pacf()*, that the partial autocorrelation plot produces similar results to your table for the hypothesis testing part of the table. Explain what you are looking to in your table and the *pacf ()* plot.

10. Show that for MA(2), for all k, $|R_k| < 1$ for any choice of b_1 and b_2. In other words, it is always "well behaved."

11. Consider the MA(2) model $\varepsilon_j = -b_2w_{j-2} - b_1w_{j-1} + w_j$ with *invertibility* conditions (i) $b_1 + b_2 < 1$, (ii) $b_2 - b_1 < 1$, and (iii) $|b_2| < 1$.

Pick choices of b_1 and b_2 that violate each condition, but only one condition at a time. In each case, show $b(B) = 0$ has one or more roots inside the unit circle. Since $b(B) = 0$ is a quadratic, it should be possible to do this easily.

16

MODELING PHILOSOPHY AND COMPLETE EXAMPLES

16.1 MODELING OVERVIEW

16.1.1 The Algorithm

It is now possible to fit a complete model to a set of data. The steps are as follows:

(i) Fit a saturated model, for the signal, to the data.

(ii) Use the residuals from this model to select an AR(m) model and values $\hat{a}_1, \ldots, \hat{a}_m$; the order, m, is determined by AIC or hypothesis testing.

(iii) Form a set of candidate models to fit the signal.

(iv) Fit all of the models using a filter based on $\hat{a}_1, \ldots, \hat{a}_m$ from the saturated model.

(v) Use model selection, on the filtered scale, to select the best model.

(vi) Assess the quality of the model using data splitting, R^2, R^2_{pred}, and/or sensitivity analysis.

(vii) Use the resulting model to perform whatever analysis is required—hypothesis tests, confidence intervals, prediction, etc.

16.1.2 The Underlying Assumption

The fundamental assumption for this modeling approach is that the true noise is very complex, perhaps more complex than even the most general ARMA(m,l) model. Nevertheless, for any finite sample, some AR(m) model fits the noise reasonably

Basic Data Analysis for Time Series with R, First Edition. DeWayne R. Derryberry.
© 2014 John Wiley & Sons, Inc. Published 2014 by John Wiley & Sons, Inc.

well. Because there is no actual belief that an AR(m) model is correct, AIC rather than BIC will be used to select the best model. The main motivations for this are that : (i) it is easy to develop an AR(m) filter, but it is difficult to develop a similar tool for general ARMA(m,l) models; (ii) a rationale has been given for the view that ARMA(m,l) = AR(∞) [i.e., $g(B)^{-1}\varepsilon_n = w_n$]; and (iii) for large data sets and realistic modeling problems, it must be the case that AR(∞) $\approx$ AR(m), where $m \ll n$.

Is this assumption valid? The assumption can neither be proven nor disproven mathematically. Even if the arguments of the last paragraph could be made quite rigorous, they would probably only apply as $n \to \infty$. The data sets in this book are all small relative to ∞! This assumption fairs well for the data sets used in this book and in the author's limited experience. The validity of this assumption, in practice, may be quite narrow or quite broad and may vary from application to application, so that some experts might be quite satisfied with the models generated using this assumption, while others might find this approach generally unreliable. This is a modeling approach that works in many situations.

But how would one even know that it works? What does that mean? Remember the goal. This approach works when the filtering process produces observations that are quasi-independent without substantially reducing the sample size. In the examples in this book, the purpose is not to model the noise precisely or explicitly. The goal is to produce "modified" data that can be analyzed using standard statistical tools that assume independence and produce results that can still be used to answer questions posed by the original, "unmodified" data.

16.1.3   An Example Using an AR(m) Filter to Model MA(3)

Perhaps the greatest challenge to this assumption would involve modeling pure MA(l) models with AR(m) models. In this example, errors from an MA(3) model are generated. The function $ar.yw()$ is used to find an appropriate order, m, for an AR(m). Finally, the errors are filtered using this model and the resulting filtered errors appear much like white noise. Of course, no AR(m) filter can really convert MA(3) errors to white noise, but the approach seems to be good enough, which is the point. To say the algorithm works [based on the assumption all noise can be modeled as AR(m), for some, perhaps large, m] means that the filtering process produces quasi-independent observations.

Consider an MA(3) model with $n = 200, 500, 5000$. Is it realistic to fit an AR(m) model? In the best-case scenario, it is expected that, as sample size increases, AIC will select models of higher and higher order for AR(m), but the order grows much more slowly than the sample size. The resulting filtered series will look a lot like white noise (remember, it is impossible to prove anything is white noise), even though increased sample sizes make it easier to detect smaller deviations from white noise. Recall, however, that an individual simulation can only be used to argue that an idea is plausible.

$$\text{Let } b(B) = (1 - 0.6B)^2(1 + 0.4B) = 1 - 0.8B - 0.12B^2 + 0.144B^3$$

```
error_500 <- arima.sim(n=500, list(ma=c(0.8, 0.12, -0.144)),sd = 1.0)
```

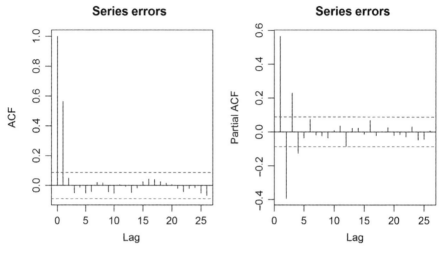

FIGURE 16.1   Errors from an MA(3) model, with $n = 200$.

The *acf( )* and *pacf( )* plots for this simulated data are displayed in Figure 16.1.
z <- ar.yw(error) and z$ar produce the estimates $\hat{a}_1 \dots \hat{a}_7$: [1] 0.85166969

-0.48968235	0.22753267	-0.05211647	-0.16460158	0.27518960
-0.15182226				

The model fitted is AR(7). The filter, a trivial generalization of the AR(1) filtering
matrices constructed in Chapter 13, is constructed as follows:

```
n <- 200                        # sample size
m <- 7                          # order of AR(m) model
tmp <- rep(0,(n-m)*n)
A <- matrix(tmp, n-m,n)
for(j in 1:(n-7))
{ A[j,j] <- 1
A[j,j+1] <- -0.85166969
A[j,j+2] <- 0.48968235
A[j,j+3] <- -0.22753267
A[j,j+4] <- 0.05211647
A[j,j+5] <- 0.16460158
A[j,j+6] <- -0.2751896
A[j,j+7] <- 0.15182226}
filtered_error <- A%*%error          #filtered errors
ar.yw(filtered_error)       Order selected 0  sigma^2 estimated as  0.843.
```

The *pacf( )* plot for the filtered error, Figure 16.2, shows no obvious evidence on
deviations from white noise.

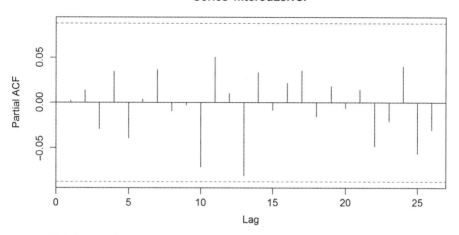

FIGURE 16.2   Filtered MA(3) errors, using an AR(7) filter, when $n = 200$.

Continuing with the same model and $n = 500$, ar.yw(error_500), this time the errors are diagnosed, by *ar.yw()* using AIC, as order 5. The *pacf()* plot is consistent with this diagnosis (Figure 16.3, left panel): Order selected 3 sigma^2 estimated as 0.9971.

Filtering as before produces ar.yw(filtered_error) Order selected 0 sigma^2 estimated as 0.9928. The *pacf()* plot in the right panel of Figure 16.3 reflects this.

This is probably a good time to mention that, in general, as sample size increases, the order, m, for AR(m) increases, but the data does not always cooperate. Recall that this is because the true model for the noise may well be AR(∞), and larger sample

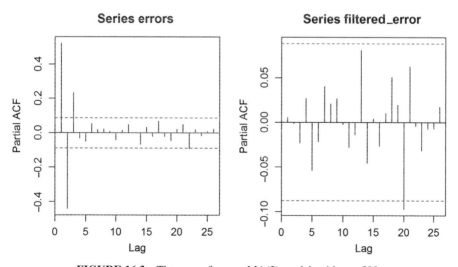

FIGURE 16.3   The errors from an MA(3) model, with $n = 500$.

sizes allow for picking up more terms, $\hat{a}_m^m$, as distinguishable from white noise. The case with $n = 5000$ is left as an exercise.

16.1.4 Generalizing the "Skip" Method

In Section 13.7.3, the "skip" method was introduced with the argument that, for AR(1) models, taking observations k units apart, for some sufficiently large k, will produce quasi-independent observations (the premise being that the data set is sufficiently large that floor(n/k) observations are still a large data set).

If the grand assumption of this chapter is true, the basis for the "skip" method is more general. Suppose a large series of observations is ARMA(m,l). It has already been supposed that there exists a p such that ARMA(m,l) $\approx$ AR(p). However, as observations are taken further apart, the serial correlation decreases in intensity and the sample size gets smaller. Both of these factors are expected to cause the estimated order, p, to drop in size. Eventually the order is approximated by AR(1), then AR(0).

Consider an example. An ARMA(2,2) model was used to simulate $n = 5000$, observations (y), and the skip method of Section 13.7.3 applied. The R code y <- arima.sim(list(ar=c(0.7, -.25), ma = c(0.7,-0.35)),n = 5000) produced the data displayed in Figure 16.4.

Since the skip method finds a skip distance for AR(1) models, the function $ar.yw()$ can be forced to fit an AR(1) model: z <- ar.yw(y, aic=FALSE, order.max=1).

z$ar 0.6141627. Applying the formula from 13.7.3 with c = 1.05 produces

$$\log_e[(0.1025)/(2.1025)]/\log_e(.6141627) = 6.197.$$

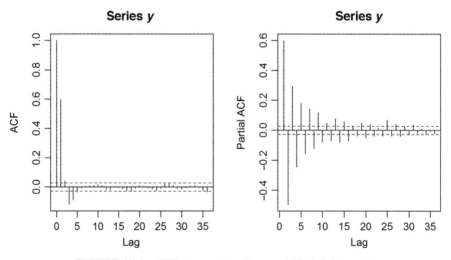

FIGURE 16.4 5000 observations from an ARMA(2,2) model.

TABLE 16.1   The "Skip" Method Applied to Some ARMA(2,2) Observations. The Third Column was Found Using ar.yw(?, aic=F,order.max= 1)

Skip step	Estimated order ar.yw()	Estimated a_1^1	Estimated variance ar.yw()	$\hat{\sigma}^2$	Ratio
$k = 1$	33	0.6142	1.104	3.187	2.89
2	2	0.0701	3.115	3.181	1.02
3	1	−0.1108	3.075	3.112	1.01
4	1	−0.1244	3.057	3.103	1.02
5	1	−0.0495	3.165	3.170	1.00
6	0	0.0073	3.193	3.193	1.00
7	3	−0.0073	3.095	3.112	1.01
8	0	−0.0038	2.95	2.95	1.00

Consider the serial correlation between observations k units apart for this series. Table 16.1 displays the impact of skipping different numbers of units. Because most statistical analyses are based on the size of variances, the skip method is effective when the variances with and without filtering are about the same. The ratio of these variances is the last column in Table 16.1.

Perhaps the most noticeable thing about the skip method, for this particular data, is that huge gains are made just by skipping to every other observation. As soon as any skipping is done, these numbers are quite comparable, and little harm is done, from the point of view of the width of confidence intervals or the size of p-values, by assuming independence. However, at a skip of 6 or 8 units, the errors are no longer distinguishable from white noise.

So, in this case, just skipping with $k = 2$, leaving 2500 observations, seems quite adequate. Certainly, skipping 6 is fully adequate (with 833 observations), and skipping 7, as recommended by the method, seems to pick up some idiosyncratic aspect of the data, producing an apparent order 3 pattern. However, even in this odd case, the ratios of the variances with and without filtering are essentially the same (ratio = 1.01).

While most of the autocorrelation is removed by skipping just every other observation, the *pacf()* plot, Figure 16.5, confirms Table 16.1 that skipping to every sixth observation is the first place where the data appears quasi-independent.

16.2   A COMPLEX PERIODIC MODEL—MONTHLY RIVER FLOWS, FURNAS 1931–1978

16.2.1   The Data

The monthly river flow (in cm) for Furnas ("Furnas 31 78.txt" from the Exercises folder) is a straightforwardly periodic collection of observations. The data clearly benefits from a logarithmic transformation (Figure 16.6, right versus left panel).

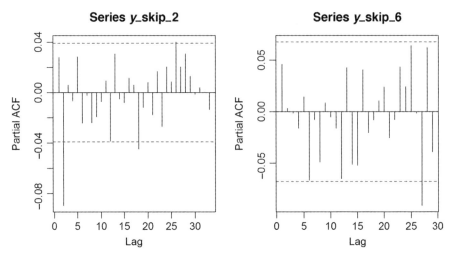

FIGURE 16.5   The plots showing the level of serial correlation for skipping 2 and 6 units for an ARMA(2,2) model.

16.2.2   A Saturated Model

In the case of annual periodic data, the saturated model is unusually straightforward. If the data is monthly, a saturated model has 12 monthly averages ($p = 12$). If the data is daily, the saturated model has 365 daily averages ($p = 365$).

The saturated model, for the monthly data, can be found using a Fourier series with the usual six pairs of trigonometric functions. The saturated model for the first 10 years of data and the *pacf()* plot for all of the data is shown in Figure 16.7.

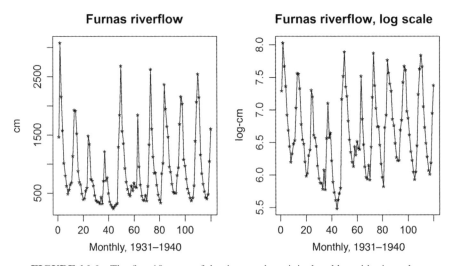

FIGURE 16.6   The first 10 years of the data on the original and logarithmic scales.

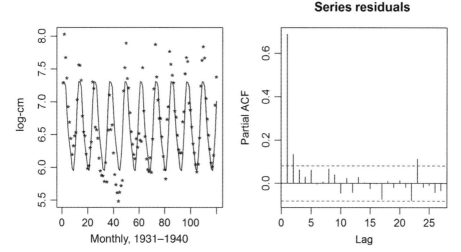

FIGURE 16.7   The fit for the saturated model.

The model will not fit the data particularly well (as the saturated model is the best possible fit). The function *ar.yw( )* estimated the order of the residuals as AR(3).

16.2.3   Building an AR(*m*) Filtering Matrix

16.2.3.1   The Linear Algebra of Filtering   Filtering is a generalization of least squares, and a number of techniques and computations will be introduced here and used throughout the book. The objective of ordinary least squares (OLS) is to find β's such that the total squared distance between $X\beta$ and y is as small as possible: min $\|X\beta - y\|$. When $\hat{\beta}$ (the value of β that solves this optimization problem) is found, then $X\hat{\beta} = \hat{y}$ are the predicted values and the vector $y - \hat{y} = \hat{\varepsilon}$ are the residuals. Furthermore, $\hat{\sigma}(X'X)^{-1}$, where $\hat{\sigma} = \text{SSE}/(n - p)$, is the variance–covariance matrix for the vector $\hat{\beta}$. Also, $\hat{\beta} = (X'X)^{-1}X'y$ (Neter et al., 1989, p. 242). All these work well when the y observations are independent (at least approximately), and the solutions have various optimal statistical properties. However, when the observations are not independent, things go awry.

The filter, A, produces linear combinations of the original vector y, such that $Ay = y_F$ are quasi-independent. The new problem min $\|AX\beta_F - Ay\| \equiv$ min $\|X_F\beta_F - y_F\|$ has the correct properties and $X\hat{\beta}_F = \hat{y}_F$ are the predicted values, and $Ay - A\hat{y} = y_F - \hat{y}_F = A\hat{\varepsilon} = \hat{\varepsilon}_F$ are the residuals for the filtered model. Furthermore, $\hat{\sigma}_F(X'A'AX)^{-1}$ is the variance–covariance matrix, where $\hat{\sigma}_F = \text{SSE}_F/(n - m - p)$. Given all this, it should come as no surprise that $\hat{\beta}_F = (X'A'AX)^{-1}X'A'Ay$.

If $A = I$, no filtering occurs and all the formulas reduce to OLS. If $A = W$, a diagonal matrix with positive, but unequal values on the diagonal, is weighted least squares and adjusts for unequal variance when observations are independent.

16.2.3.2 Construct of a Matrix A and Fitting a Model with lm() Suppose a vector y has n observations and it is determined that the filter is of order m. The R code for every filtering matrix, A, used in this book is simply:

```
tmp <- rep(0, (n-m)*n)
A <- matrix(tmp, n-m, n)
for(j in 1:(n-m))
{A[j,j] <- 1
A[j, j+1] <- -â₁
...      A[j,j+m] <-âₘ }
```

This is a generalization of the AR(1) filter constructed in Section 13.6.3.2. A complete matrix, for a specific case, was already presented in Section 16.1.3.

A model is fitted with a command like fit_filtered <- lm(A%*%y~-1+A%*%X).

It is implicit in this formulation that the first column of X is a column of ones, so that the intercept is also filtered. Unless this is done, the intercept will not be estimated properly (occasionally this is important).

16.2.3.3 Standard Computations Statistical analysis occurs for the filtered model, where the optimal statistical properties reside; interpretation occurs in the unfiltered model, where humans reside. This will be a common theme throughout the rest of the book.

For model selection, AIC_F, R_F^2, and $R_{F,\text{pred}}^2$ are important (because model selection is on the filtered scale). These are, of course, just the usual suspects, but computed on the filtered data.

These are all built up from

$$n_F = n - m, \text{SST}_F = \text{var}(y_F)(n_F - 1), \text{SSE}_F = \sum \hat{\varepsilon}_{F,j}^2, \text{and}$$

$$\text{PRESS}_F = \sum (\hat{\varepsilon}_{F,j}/[1 - h_{F,jj}])^2,$$

where the extraction of the "hat" values, $h_{F,jj}$, is discussed in Section 11.6.3. If the observations have about the same influence, a common situation in time series, then $\text{PRESS}_F \approx \text{SSE}_F(n_F/[n_F - p])^2$. So

$$R_F^2 = 1 - \text{SSE}_F/\text{SST}_F, R_{F,\text{pred}}^2 = 1 - \text{PRESS}_F/\text{SST}_F \text{ and}$$

$$\text{AIC}_F = n_F \cdot \log_e(\text{SSE}_F) + 2(p + 1).$$

For interpretation on the original scale, both R^2 and R_{pred}^2 are useful. All the formulas above hold with the subscript F removed. The only modification is that the "hat" values from an OLS problem are used in the computation of PRESS.

In general, $R_{\text{OLS}}^2 \geq R^2 \geq R_{\text{prod}}^2$, where R_{OLS}^2 is the explained variation from the OLS fit.

16.2.3.4   The Filtering Information for This Model   From *ar.yw()* for the residuals from the saturated model:

0.5877 0.0961 0.0629   Order selected 3 sigmaˆ2 estimated as 0.06456.

16.2.4   Model Selection

A number of models will be fitted to all of the data (the first few years appear in graphs because plotting all of the data produces very "busy" displays) on the logarithmic scale. The most complex model to be considered will have six pairs of trigonometric functions, the simplest will have just one pair. All fitting and model selection will occur on the filtered scale. However, there is just one filtering matrix, A, which is based on the residuals from the saturated model.

Given that there are 576 observations, it is not difficult to believe that the best model (Table 16.2) has close to 12 parameters, the most possible if it is assumed that the data follows a purely periodic pattern.

A model with five pairs of trigonometric functions and a mean can be expected to an R^2 of about 64.30% (Table 16.2) for future data, on the original scale. As models become greatly oversimplified, biases are introduced into the residuals. All of the models with five or more parameters have residuals that are diagnosed as white noise, using *ar.yw()*. However, the three-parameter model appears AR(6) and the single mean model appears AR(25) for the resulting residuals.

TABLE 16.2   Model Selection Information for the Furnas 31.78 Data ($n = 573$), with AR(3) Filter

Model	p	R^2_F (R^2)	SSE_F	AIC_F	$R^2_{F,pred}$ (R^2_{pred})
Single mean	1	NA	92.932 $(= SST_F)$	2600.760	NA
One pair	3	0.5528 0.6366	41.560	2143.652	0.5481 0.6327
Two pairs	5	0.5923 0.6539	37.889	2094.657	0.5851 0.6478
3	7	0.5946 0.6546	37.671	2095.35	0.5845 0.6460
4	9	0.5998 0.6556	37.190	2091.996	0.5869 0.6448
5	11	0.6035 0.6565	36.850	2090.728[a]	0.5878[a] 0.6430
6	12	0.6042 0.6566	36.780	2091.633	0.5856 0.6405

[a]Best model.

16.2.5   Predictions and Prediction Intervals for an AR(3) Model

Predictions for AR(m) are straightforward generalizations of the AR(1) approach, although the algebra is quite messy. Suppose the first n errors, $\varepsilon_1, \ldots \varepsilon_n$, the values $a_1 \ldots a_m$, the signal, and the white noise variance, σ_w^2 are all known and the goal is to predict the next k observations including prediction intervals. (In practice, all of these values are replaced with estimates, producing relatively accurate, but approximate, intervals.) The first three predicted future errors for an AR(3) model are computed, although the pattern for any future prediction is straightforward, but tedious (however, it should be easy to program using iterative programming structures). These formulas are based on the definition of an AR(m) model, a clear understanding that errors $\varepsilon_1, \ldots \varepsilon_n$ have been realized, but that errors $\varepsilon_{n+1}, \ldots \varepsilon_{n+k}$ as random variables, and that iterative substitution can be used to reduce all future ε_{n+k} to a linear combination of the last m known errors and future white noise:

$$\hat{\varepsilon}_{n+1} = a_1 \varepsilon_n + a_2 \varepsilon_{n-1} + a_3 \varepsilon_{n-2} + w_{n+1}$$

$$\hat{\varepsilon}_{n+2} = a_1 \hat{\varepsilon}_{n+1} + a_2 \varepsilon_n + a_3 \varepsilon_{n-1} + w_{n+2}$$

$$= (a_1 + a_2)\varepsilon_n + (a_1 a_2 + a_3)\varepsilon_{n-1} + a_1 a_3 \varepsilon_{n-2} + a_1 w_{n+1} + w_{n+2}$$

$$\hat{\varepsilon}_{n+3} = a_1 \hat{\varepsilon}_{n+2} + a_2 \hat{\varepsilon}_{n+1} + a_3 \varepsilon_n + w_{n+3}$$

$$= \left(a_1^2 + 2a_1 a_2 + a_3\right) \varepsilon_n + \left(a_1^2 a_2 + a_1 a_3 + a_2^2\right) \varepsilon_{n-1} + \left(a_1^2 a_3 + a_2\right) \varepsilon_{n-2}$$

$$+ \left(a_1^2 + a_2\right) w_{n+1} + a_1 w_{n+2} + w_{n+3}.$$

In other words, the predicted errors, attached to future observations, are dependent on the last three known errors [hence an AR(3) model]. For the real world approximation, all future predicted errors will be based on the last three reported residuals.

The means and variances are quite straightforward:

$$E(\hat{\varepsilon}_{n+1}) = a_1 \varepsilon_n + a_2 \varepsilon_{n-1} + a_3 \varepsilon_{n-2}; \quad \text{Var}(\hat{\varepsilon}_{n+1}) = \sigma_w^2;$$

$$E(\hat{\varepsilon}_{n+2}) = (a_1 + a_2)\varepsilon_n + (a_1 a_2 + a_3)\varepsilon_{n-1} + a_1 a_3 \varepsilon_{n-2}; \quad \text{Var}(\hat{\varepsilon}_{n+2}) = \left(a_1^2 + 1\right)\sigma_w^2;$$

$$E(\hat{\varepsilon}_{n+3}) = \left(a_1^2 + 2a_1 a_2 + a_3\right) \varepsilon_n + \left(a_1^2 a_2 + a_1 a_3 + a_2^2\right) \varepsilon_{n-1} + \left(a_1^2 a_3 + a_2\right) \varepsilon_{n-2};$$

$$\text{Var}(\hat{\varepsilon}_{n+3}) = \left(\left[a_1^2 + a_2\right]^2 + a_1^2 + 1\right)\sigma_w^2.$$

The predictions are then $\hat{y}_{AR,n+k} = \hat{y}_{n+k} + \hat{\varepsilon}_{n+k}$, where $\hat{y}_{n+k}$ is the prediction from the signal, $\hat{\varepsilon}_{n+k}$ is the predicted error, and $\hat{y}_{AR,n+k}$ is a prediction from the signal corrected for the AR(m) structure of the model. An approximate prediction interval is $\hat{y}_{AR,n+k} = \hat{y}_{n+k} + \hat{\varepsilon}_{n+k} \pm 2\sqrt{\text{Var}(\hat{\varepsilon}_{n+k})}$. Intervals computed using software may have additional adjustments for replacing parameters with estimates and will be slightly wider.

16.2.6 Data Splitting

16.2.6.1 An Overview   Data splitting is the best way to assess predictive accuracy, and this data set is large enough for data splitting. A training set of 32 years and a validation set of 16 years follows previously mentioned guidelines. Because the errors follow an AR(m) pattern, both the training set and the validation set should be contiguous values. The data will be fitted using the last 32 years, while validation will involve assessing how well this model fits the first 16 years. The filter will be built only on a saturated model for the training set, and not all of the data is part of the uncertainly that should be captured by the data-splitting process.

The steps are:

(i)   Create a training set.
(ii) Build the saturated model for the training set.
(iii) Build filters A_t for the training set and A_v for the validation set based on the residuals from the training set saturated model.
(iv) Fit all models with $R^2_{F,\text{cross}}$, the explained variation for the validation set on the filtered data, as the criteria for selection.

16.2.6.2 Some Linear Algebra   The procedure now has six major components: X_t, A_t, and y_t for the training set and X_v, A_v, and y_v for the validation set. The mathematics of fitting a training set is exactly as already outlined. The goal is to solve the filtered problem min $\|A_t X_t \beta_{Ft} - A_t y_t\|$. This will produce predictions of the form $X_t \hat{\beta}_{Ft} = \hat{y}_t$. This is all as described in the previous section with the "t" subscripts added.

The filtering matrices, A_t and A_v, are constructed in the usual way except that the number of rows should reflected the number of observations in the training and validation sets. The filtering matrices will use the following information from *ar.yw( )* on the residuals from the saturated model on the training set: 0.5856 0.1133 0.0758   Order selected 3 sigma^2 estimated as 0.06615. The training set has residuals very similar to the original model with all of the data.

The validation process involves using these values for $\hat{\beta}_{Ft}$ to assess the fit on both the filtered and original scale for the validation set. In this case, $A_v X_v \hat{\beta}_{Ft} = A_v \hat{y}_v = \hat{y}_{F,v}$ and $X_v \hat{\beta}_{Ft} = \hat{y}_v$ are the predicted values for the validation set, on the filtered and unfiltered scale, respectively. Similarly, the residuals are $\hat{\varepsilon}_{F,v} = A_v(y_v - \hat{y}_v) = y_{F,v} - \hat{y}_{F,v}$ and $\hat{\varepsilon}_v = y_v - \hat{y}_v$. As always, computations on the filtered scale are for model selection, while computations on the original scale are for interpretation.

16.2.6.3 Data Splitting Computations   The purpose now is to compare the values R^2_t, $R^2_{F,t}$, $R^2_{t,\text{pred}}$, and $R^2_{F,t,\text{pred}}$—the measures of fit on the validation set, R^2_v, $R^2_{F,v}$. It would almost always be the case that $R^2_t > R^2_v$ and $R^2_{F,t} > R^2_{F,v}$ and it is expected that R^2_v and $R^2_{F,v}$ are better measures of future fit than $R^2_{t,\text{pred}}$ and $R^2_{F,t,\text{pred}}$, although there is an interest in the comparability of these two different approaches to cross validation.

The values R_t^2, $R_{F,t}^2$, $R_{t,pred}^2$, and $R_{F,t,pred}^2$ are all computed as before. The number AIC_F is not computed because $R_{F,v}^2$ will be the basis for model selection. Let n_v be the number of observations in the validation set.

Then $n_{F,v} = n_v - m$, $SST_{F,v} = Var(y_{F,v})(n_{F,v} - 1)$, and $SSE_{F,v} = \sum \hat{\epsilon}_{F,v}^2$. So $R_{F,v}^2 = 1 - SSE_{F,v}/SST_{F,v}$. Dropping the "F" subscript from all values, the same formulas are used to produce $R_v^2 = 1 - SSE_v/SST_v$.

16.2.7 Model Selection Based on a Validation Set

16.2.7.1 The Model Selection As before without cross validation, Table 16.3 indicates that the best model with cross validation has $p = 11$; this model was chosen because it had the largest value for $R_{F,v}^2$. The only model that fits the data noticeably worse is the simple periodic model. For all of the models, except the validation set with $p = 3$, the residuals were indistinguishable from white noise using *ar.yw()*. This suggests that the results are not highly dependent on the choice for filtering matrix, and that a single filtering matrix, based on a saturated model, often works well as a filter for all of the best candidate models.

How good should future predictions be? Using the analysis from Table 16.2 and the complete data set, $R_{pred}^2 = 0.6430$, on the training set $R_{pred}^2 = 0.6085$, and using the validation set $R_v^2 = 0.7070$. These numbers are all efforts to quantify the reliability of future predictions, and it seems clear the data fits the validation set better than the training set. This, as already iterated, does not usually happen (except in this book, where it seems to happen all the time).

It is clear that the filtered residuals from the validation set (right panel) of Figure 16.8 are indistinguishable from white noise. This is consistent with the *ar.yw()*

TABLE 16.3 Model Selection Information for the Furnas 31.78 Data, with AR(3) Filter Using Cross Validation

Model	p	$R_{F,t}^2$ (R_t^2)	$R_{F,pred,t}^2$ ($R_{pred,t}^2$)	$R_{F,v}^2$ (R_v^2)
One pair	3	0.5534	0.5463	0.5806
		0.6119	0.6057	0.6864
Two pairs	5	0.5921	0.5811	0.6280
		0.6280	0.6181	0.7034
3	7	0.5980	0.5828	0.6185
		0.6298	0.6160	0.7034
4	9	0.6000	0.5803	0.6278
		0.6303	0.6123	0.7057
5	11	0.6018	0.5777	0.6334[a]
		0.6306	0.6085	0.7070
6	12	0.6039	0.5753	0.6297
		0.6309	0.6045	0.7065

[a]Best model.

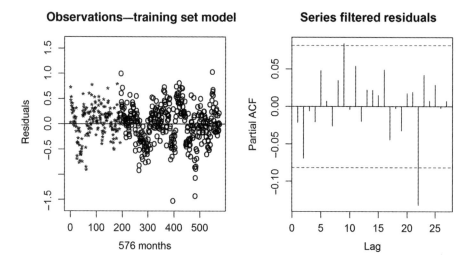

FIGURE 16.8   The filtered residuals for the validation set and the $p = 11$ model on the left. The residual from a fit of all the data to the training set model on the right.

analysis of these residuals. On the left panel of Figure 16.8, the model fit from the training set has been fitted to all of the data—the validation set residuals are represented by "*" and the training set residuals by "o." Notice that there are several outliers on the right two-thirds of the display in the training set, so it is not at all surprising that the data fits the validation set better than the training set.

16.3   A MODELING EXAMPLE—TREND AND PERIODICITY: CO_2 LEVELS AT MAUNA LAU

16.3.1   The Saturated Model and Filter

The Mauna Lau CO2 data (Figures—3.4, "MLCO2.txt" from the Examples folder) displays both periodicity and trend (perhaps nonlinear). Based on the display of the data, the periodicity does not appear to be overly complex, a model of the form $y(j) = \mu + \beta_1 j + \ldots \beta_4 j^4 + B \cdot \cos(2\pi j/k) + C \cdot \sin(2\pi j/k) + \ldots H \cdot \cos(8\pi j/k) + I \cdot \sin(8\pi j/k)$ can be taken as the saturated model. The data is monthly, so k is taken to be 12. The model is fitted by forming the usual matrix and solving using $lm()$ in R.

Why is the saturated model so complex? The residuals from the saturated model will be used to identify the order $AR(m)$, for all filtering. If the signal chosen for the saturated model is too simple, some signal will still be in the noise, and the apparent noise will have hidden bias. The noise, for the initial diagnosis of order $AR(m)$, needs to be as free of signal as possible. The signal has two components, trend and periodicity. The saturated model, provided there is enough data to build a sufficiently complex model, should be more complex than the model eventually selected.

In fact, the saturated model was chosen by adding pairs of trigonometric functions and polynomial terms until some terms became highly nonsignificant in the fitted model. In general, if there is any doubt, the saturated model should be over-fitted to the data. The only concern is that a saturated model becomes so big that the residuals are distorted. A saturated model should have $p < n/10$. Finding a good saturated model is therefore most challenging when (i) the signal is complex and the sample size is small or (ii) when a local smoother is being fitted to the data, as these models potentially have much greater complexity than models with Fourier series and polynomial terms.

In, what might seem like madness, five pairs of trigonometric terms and powers of j ($j = 1,2,3,\ldots,384$), through the fourth power, were fitted in the saturated model. This model should have the "best" possible residuals for fitting an AR(m) model.

Numerical difficulties can occur when a matrix is formed by forming columns of differing powers of a variable. Such matrices tend to have colinearity (Neter et al., 1989, p. 316), which is a bad thing. The simple transformation t <- j − mean(j) substantially reduces this problem. All polynomial terms in the model will be based on this "demeaned" variable for time.

Analysis of the residuals using ar.yw(saturated model residuals):

0.6209 0.1736   Order selected 2 sigma^2 estimated as 0.06973.

The residuals are not bad, just AR(2). Only two observations will be lost to filtering.

There is sometimes guesswork in generating the saturated model. If all of the terms in the current saturated model were significant, which they are not (Table 16.4), a more complex model would clearly be required as the saturated model. It is also useful to keep the trend and periodic components of the model clearly segregated (Table 16.4).

The residuals from the saturated model display clear AR(2) behavior (Figure 16.9).

16.3.2 Model Selection

16.3.2.1 An Overview   One approach to convergence on the "best" model is to drop variables that seem unimportant and keep dropping them as long as it is possible to improve AIC_F or $R^2_{F,pred}$. Then choose the model that attained the best value. In this context p-values could be used, but not as hypothesis tests. Large p-values (or small t-ratios) indicate which variables might be dropped with little impact on the current fit of the model to the data. This approach produces Table 16.5.

It is important to distinguish two apparently similar, but very different, methods of model selection based on p-values. A common approach would be to drop eligible variables with large p-values until a model is found in which all variables eligible to drop have p-values less than some cut-off (almost always that supposedly "magical" 0.05, of course).

TABLE 16.4   The Saturated Model Fitted to the Mauna Lau CO_2 Data

Coefficients:

	Estimate	SE	t-value	Pr(>$|t|$)
(Intercept)	3.305e+02	3.951e–02	8366.061	< 2e–16
Polynomial terms				
Xt	1.066e–01	4.754e–04	224.264	< 2e–16
Xt_2	1.327e–04	6.710e–06	19.771	< 2e–16
Xt_3	–1.246e–07	1.971e–08	–6.320	7.56e–10
Xt_4	1.880e–11	2.035e–10	0.092	0.926460
Periodic terms				
Xcol_c1	–1.103e+00	2.980e–02	–37.018	< 2e–16
Xcol_s1	2.490e+00	2.983e–02	83.471	< 2e–16
Xcol_c2	6.564e–01	2.980e–02	22.027	< 2e–16
Xcol_s2	–3.668e–01	2.981e–02	–12.307	< 2e–16
Xcol_c3	4.319e–02	2.980e–02	1.450	0.148032
Xcol_s3	–9.900e–02	2.980e–02	–3.322	0.000984
Xcol_c4	–7.525e–02	2.980e–02	–2.525	0.011983
Xcol_s4	4.520e–02	2.980e–02	1.517	0.130230
Xcol_c5	–2.117e–02	2.980e–02	–0.710	0.477853
Xcol_s5	1.339e–02	2.980e–02	0.449	0.653477
Residual SE: 0.4129 on 369 df.				

Another approach is to drop an eligible variable with the largest current p-value (however large or small), attaching a model selection value (for example, AIC or R^2_{pred}) to each model. This approach continues to drop variables, even if they have small p-values, until the model selection value can no longer be improved. In this method, the model that attained the best value for the model selection value is chosen.

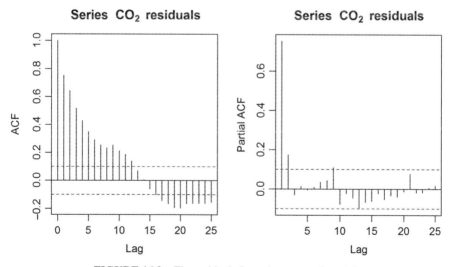

FIGURE 16.9   The residuals from the saturated model.

TABLE 16.5   Model Selection Information for the Filtered Mauna Lau CO_2 Data ($n_F = 382$). The Model is or $= r$ (Order of Polynomial, Including Intercept), and pr $= s$ (Number of Fourier Series Pairs)

Model	$p = r + 2s$	R_F^2 (R^2)	SSE_F	AIC_F	$R_{F,pred}^2$ (R_{pred}^2)
Saturated	15	0.9904	26.44	1282.96	0.9896
or $= 5$, pr $= 5$		0.9988			0.9987
or $= 4$, pr $= 5$	14	0.9904	26.44	1280.96	0.9896
		0.9988			0.9987
or $= 4$, pr $= 4$	12	0.9903	26.71	1280.87[a]	0.9897[a]
		0.9988			0.9987
or $= 3$, pr $= 4$	11	0.9902	26.91	1281.73	0.9896
		0.9987			0.9986
or $= 3$, pr $= 3$	9	0.9891	30.05	1319.91	0.9886
		0.9986			0.9986
or $= 2$, pr $= 1$	4	0.9330	184.40	2002.93	0.9316
		0.9808			0.9803
Single mean	1	NA	2753.76	3029.72	NA

[a]Best model.

The latter is being used here, but it is easy for neophytes to confuse the two procedures, which often end up with the same resulting best model.

16.3.2.2   *Pruning a Saturated Model*

The model is pruned as long as progress is possible with regard to AIC or R_{pred}^2. With regard to the polynomial terms, it makes no sense, for example, to drop a quadratic term, when the cubic term is still in the model, so only the highest power is eligible to be dropped at any point in time. Similarly, but perhaps less obviously, the only terms eligible to be dropped from the periodic component is the pair with the highest frequency, and both the sine and cosine function are dropped together.

In other words, this model is highly structured and at any iteration of the model selection process, either the higher order polynomial or the last trigonometric pair is the only term eligible to be dropped. All models with $p \geq 11$ had residuals not significantly different from white noise, based on $ar.yw()$.

Model selection does not stop when all the p-values for the remaining terms are small. Model selection ends when no further progress can be made. It may not be obvious, but in Table 16.5, once the quadratic model with three pairs of trigonometric functions is fitted to the data, it is clear that no further exploration can produce a model better than the cubic model with four pairs of sine and cosine functions.

Why? First, notice that AIC_F cannot be improved. Any model that is simpler than the quadratic three-pair will have an SSE of at least 30.05. This means any such model will have an AIC_F of at least $382 \cdot \log_e (30.05) + = 1299.89+$. It is just as obvious that $R^2_{F,pred}$ cannot be improved. No model simpler than the cubic three-pair model can have an R^2_F better than 0.9891, furthermore, it is always the case $R^2_{F,pred} \leq R^2_F$, so not simpler model has the prospect of improving $R^2_{F,pred}$. Both criteria select the same best model and suggest that the three middle models are all quite close.

Although there is always one best model, several all seems to fit the data about equally well. Notice that even a very simple model, a linear trend combined with a simple periodic function (or $= 2$, pr $= 1$), produces a relatively high-quality fit on the original scale.

16.3.3   How Well Does the Model Fit the Data?

The model has been fitted on the filtered scale, but what does the model look like on the original scale? The model has both a trend and a periodic component; it is natural to break the fit up into these components. While the numerical information is important, graphs usually tell a better story. In this case, it is important to separate the trend and periodic elements, and attempt to convey how well, or poorly, the model fits each element. This method of graphing the data versus components of the signal will be used several times in the book.

The following two fits partition the X matrix into a trend and periodic part. Each graph adds the complete model residuals to the partial fit (periodic or trend), so each graph *treats all the variations as associated with the partial fit being displayed (trend or periodic), as if the ignored aspect (periodic or trend) were deterministic.* In other words, this is a conservative representation of how well each component fits the data.

```
# How well does the trend fit the data, on the original scale?
residuals <- yCO2 - X%*%fit$coeff     #the residuals for the model
trend_fit <- X[,1:4]%*%fit$coeff[1:4]      # the model for the trend
trend_y <- trend_fit + residuals        # the trend + all of the noise
plot(1:384, trend_y,pch = "*")
lines(1:384,trend_fit)
title("The trend aspect of the fitted model")
#How well does the periodic aspect of the model fit the data
periodic_fit <- X[,5:12]%*%fit$coeff[5:12]    # the model for the period
periodic_y <- periodic_fit + residuals # the periodic fit + all of the noise
plot(265:384,periodic_y[265:384],pch = "*")
lines(265:384,periodic_fit[265:384])
title("The periodic aspect of the data, last 10 years")
```

The graph (Figure 16.10) was too busy with all 32 years of data, so the graph is for the last 10 years (for the periodic component).

The upper two graphs, based on the original scale, show a very strong fit of the model to the data. The lower parts of the graph evaluate the fit of the filtered model. The plot of the fitted values versus observed values shows a strong relationship. The

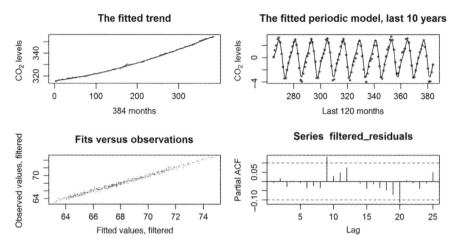

FIGURE 16.10   The fit of the best model to the Mauna Lau CO_2 data.

pacf( ) plot, confirmed using *ar.yw( )*, shows residuals not distinguishable from white noise. Overall, the model fits the data very well.

It appears that CO_2 levels vary periodically by about ± 3 parts per million (ppm) from the trend curve; the highest values being around May and the lowest values being around October.

Estimating the rate of increase per month is quite complex, due to the cubic model. Falling back to the linear model (this will tend to underestimate the rate of change because the curve appears to be concave), and the curve does look essentially linear, the CO_2 levels are increasing at 0.1023 ppm per month (12.3 ppm per decade). However, this is an extrapolation (problematic) from an observational study (problematic). To assume that this trend will continue, some scientific reason must be given to explain the cause of the trend seen in the data. Extrapolations and causality can never be based on statistical analysis alone. The scientific reasons may or may not be clear, but they are not the topic of a book on statistical analysis. Furthermore, the linear model does not have white noise residuals, so confidence intervals (as opposed to point estimates) would require the best model.

16.4   MODELING PERIODICITY WITH A POSSIBLE INTERVENTION—TWO EXAMPLES

16.4.1   The General Structure

Two examples will be given for intervention models. An intervention model generally has three components: periodicity, trend, and an abrupt change in trend at a fixed point in time. The fixed point may be known or may need to be determined.

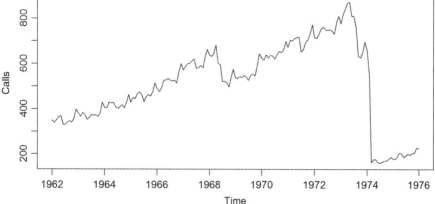

FIGURE 16.11   A drop in usage of directory assistance (Examples data folder "Directory assistance.txt" or DataMarket – Time Series Data Library – Miscellaneous – Monthly average daily calls to directory assistance January 1962–December 1976).

16.4.2   Directory Assistance

16.4.2.1   The Data   Certainly a case in which the intervention is obvious is taken from the monthly volume of directory assistance calls in a 15-year period from 1962 to 1976 (Figure 16.11).

16.4.2.2   A Question   The concern here is to estimate the drop in directory assistance usage that occurred between period 146 (February 1974) and period 147 (March 1974). Of course, the difference in usage between these two points could be used to estimate the drop as 387 ($549 - 162$) calls per month. It is unknown from what region this data was taken. The reason for the drop is not clear, but most likely it is due to change in policy from providing free directory assistance to charging for the service (based on anecdotal discussions with students).

Why might one want to estimate this number? Another region or group may want to know the implications of adding or dropping various fees for services. Getting a good estimate here would be informative in future such decisions. It is also the case that an interval estimate is preferred to a point estimate. It is worth considering how interval estimates, or any statistical estimate, gets attached to a unique event. An exercise gives the student an opportunity to think about this issue.

16.4.2.3   The Saturated Model and Filtering Information   This model will contain all the usual components associated with a saturated model: a very complex periodic function and a high-order polynomial for the trend. However, this model will also contain a shift in mean beginning with observation 147. This entails the inclusion of a column of the form: shift <- c(rep(0,146),rep(1,34)) in the X matrix. The filtered model will, then, produce a confidence interval for this mean directly.

The saturated model includes a fourth-degree polynomial, a shift term (which will be included in every model), and the usual trigonometric pairs. The residual from this model are used to develop a filter. From *ar.yw()*: 0.6988 0.1230 0.1406 -0.0385 0.0277 -0.1731 Order selected 6 sigma^2 estimated as 564.7.

In fact, this is the first example of a common theme. This data must be filtered twice. A second filter, 0.1651 Order selected 1 sigma^2 estimated as 424.4. The need to filter twice, a situation often encountered, will be discussed after the main analysis.

16.4.2.4 Model Selection—When AIC_F and $R^2_{F,pred}$ Select Different Models   The

model selection information in Table 16.6 ends with the model with $p = 10$. There are two reasons to drop from the consideration the models with $p \leq 10$. In fact, dropping additional variables beyond the $p = 12$ model is a bit contrived. From Table 16.7 it is easy to see that all the variables still in the model are highly significant.

Furthermore, when the last pair of trigonometric functions is dropped, the errors become AR(6), instead of AR(0), as indicated using *ar.yw()*. As before, when a part of the signal is dropped, and nonwhite noise errors appear suddenly, this suggests that bias has been introduced to the noise by dropping a key element of the signal. This highlights a point made throughout the book: often hypothesis testing, information criteria, and residuals analysis, all point to the same conclusion.

Returning to model selection—Table 16.6—in this case, the model selection problem is quite interesting. AIC_F and $R^2_{F,pred}$ select different models, both for the best and for the second best model. How does this happen? The nature of this model is such that observation 146 is a leveraged point and has a very high "hat" value in the filtered model. This causes the $R^2_{F,pred}$ to be substantially lower than R^2_F, and $R^2_{F,pred}$ picks models that reduce the influence of this observation (in Appendix B, it is argued

TABLE 16.6   Model Selection Information for the Filtered Directory Assistance Data ($n_F = 174$). or $= r$ (Order of Polynomial) and pr $= s$ (Number of Fourier Series Pairs). All Models Include a Shift Term

Model	p	R^2_F (R^2)	SSE_F	AIC_F	$R^2_{F,pred}$ (R^2_{pred})
Saturated:	14	0.8556	72,550.19	1966.22	0.5933
or $= 5$, pr $= 4$		0.9191			0.9017
or $= 4$, pr $= 4$	13	0.8555	72,653.26	1964.47[a]	0.5990
		0.9233			0.9093
or $= 3$, pr $= 4$	12	0.8505	75,104.98	1968.21	0.6408[a]
		0.9240			0.9121
or $= 3$, pr $= 3$	10	0.8310	84,874.87	1985.37	0.6240
		0.9216			0.9114

[a]Best model.

TABLE 16.7   The Summary of the Filtered Fit with $p = 12$

Coefficients:

	Estimate	Std. Error	t-value	Pr($>|t|$)
A2 %*% (A %*% X)mean	6.268e+02	1.337e+01	46.880	< 2e–16
A2 %*% (A %*% X)shift	−4.507e+02	2.177e+01	−20.702	< 2e–16
A2 %*% (A %*% X)t	2.086e+00	2.149e–01	9.703	< 2e–16
A2 %*% (A %*% X)t_2	−1.779e–02	4.441e–03	−4.006	9.42e–05
A2 %*% (A %*% X)col_c1	−6.415e+00	4.517e+00	−1.420	0.157515
A2 %*% (A %*% X)col_s1	−1.607e+01	4.515e+00	−3.560	0.000487
A2 %*% (A %*% X)col_c2	5.329e+00	2.047e+00	2.603	0.010093
A2 %*% (A %*% X)col_s2	−9.566e+00	2.049e+00	−4.669	6.34e–06
A2 %*% (A %*% X)col_c3	−9.059e+00	1.990e+00	−4.553	1.04e–05
A2 %*% (A %*% X)col_s3	6.942e+00	2.009e+00	3.456	0.000701
A2 %*% (A %*% X)col_c4	8.628e–01	1.415e+00	0.610	0.543029
A2 %*% (A %*% X)col_s4	−6.448e+00	1.421e+00	−4.539	1.10e–05

AIC and R^2_{pred} will pick the same model when sample size is large *and there are no influential observations*).

16.4.2.5   Using the Models to Answer the Question   Typically, the best-fitting model would be used to get an interval estimate for the shift parameter. However, if there is no clear best model, perhaps several models should be examined to see how much the interval changes, based on the model chosen. This is a form of sensitivity analysis: how sensitive to the choice of model is the interval estimate of interest.

It is possible to produce a confidence interval for the shift parameter for each of the fitted models, comparing the widths of the intervals (Table 16.8).

All of these models show similar results, especially compared to the quick and dirty calculation of 387 presented at the beginning of this section. Reports to decision makers are usually in round numbers. A range of values from 400 to 500, for example, would be broadly consistent with any and all of the intervals in Table 16.8.

Many debates and arguments could be avoided by beginning with the question— does it make much difference? In this case, reporting a slightly wider interval than those found contains the key information while avoiding a decision about which is the best model. In general, it is good to find ways to report wider intervals. The intervals developed in statistics generally do not account for all sources of uncertainty and are

TABLE 16.8   Interval Estimates for the Drop in Directory Assistance Usage in March 1974

Model	Point estimate	SE	Interval
Saturated: shift, fourth order, four pairs, "best" AIC_F	439.7	22.43	394.8–484.6
Shift, cubic, four pairs "best" $R^2_{F,pred}$	442.7	22.19	398.3–487.1
Shift, quadratic, four pairs	456.1	21.63	412.8–499.4

generally too narrow. In this case, reporting the wider interval is an informal way of accounting for modeling uncertainty—the uncertainty as to which model fits the data.

16.4.2.6 Filtering Twice   This data set was filtered twice. Why? Usually, filtering need only be applied once to get a good result. However, complications can occur, especially when the data set is small or the order of the residuals is large. Filtering is motivated by the following argument.

$y \sim X\beta$ and $A(y - X\beta) \sim$ *white noise*. That is, the filtering matrix, A, is intended to make the original results look like white noise. The least squares problem produces a second set of residuals: $Ay \sim AX\beta_F$ and $A(y - X\beta_F) \sim$ *white noise*.

Usually $\beta \approx \beta_F$, so the residuals change little and the residuals are still about as close to white noise after the second regression. However, it is possible for β and β_F to be very different. In this case, a second, or even a third filter, may be applied. When multiple filters are applied, the effective sample size, n_F, which is sometimes already small, continues to diminish. Nevertheless, if the final vector $A_3 A_2 A_1 y$ has quasi-independent observations, the results are valid on the filtered scale and produce information useful on the original scale.

There was actually an early warning that there could be trouble with this data set. Although the function *ar.yw()* diagnosed the residuals from the unfiltered saturated model as AR(6), the AIC values for AR(6) and AR(8) were close: z <- ar.yw(residuals) and z$aic[4:9] produced:

3	4	5	6	7	8
4.294	3.148	3.477	0.000	1.590	0.958.

Notice how AR(6) and AR(8) produce AIC values within a unit of each other.

In fact, even for the saturated model, the initial filter did not produce white noise on the filtered scale. It is often the case that just beginning with an AR(8) filter solves the problem, but in this case, a second filter would still be required, and filtering twice lost seven, rather than eight observations. Although it may seem troubling that such an *ad hoc* method is used for filtering, keep in mind three things: (i) there is no intrinsic interest in the filter itself, it is just the estimate of a complex nuisance parameter, (ii) once the model process begins, the filter does not change, and (iii) the objective is just to get quasi-independent observations.

16.4.3 Ozone Levels in Los Angeles

Large amounts of ozone can be produced by the burning of fossil fuels and ozone is viewed as a harmful pollutant when ozone concentrations are high. Reduction of the ozone level in the Los Angeles area is viewed as a health benefit. Did ozone levels drop in Los Angeles in the 18-year period from 1955 to 1972 (Figure 16.12)?

The data will be modeled as a complex periodic pattern with a trend. Hopefully an approximately monotonic downward trend in the data can be established.

The saturated model fitted will have a fourth-power polynomial (given that there are just 18 full periods, this is quite a complex model) and a periodic function with

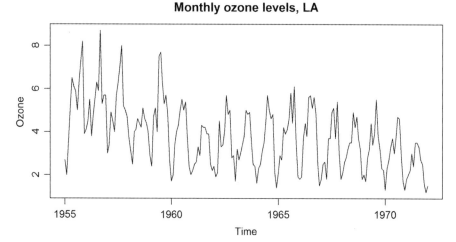

FIGURE 16.12   Monthly ozone readings ("Ozone LA.txt" from the Examples folder or DataMarket–Time Series Data library–Meteorology –Ozone concentration, downtown L. A., 1955–1972).

six pairs of trigonometric functions (this will allow each month to have a different mean).

The saturated model is just AR(1) with $\hat{a} = 0.3034$ [based on *ar.yw()*]. This is consistent with Figure 16.13.

After fitting the last model in Table 16.9, it is clear that no better model is possible with regard to either AIC_F or $R^2_{F,pred}$ (exercise). The best model has a very complex (fourth-order) polynomial structure, but a relatively simple periodic structure.

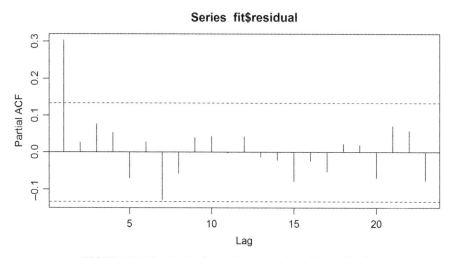

FIGURE 16.13   Residuals for the saturated model are AR(1).

TABLE 16.9   Model Selection Information for the Filtered LA Ozone Data ($n_F = 215$). or $= r$ (Order of Polynomial) and pr $= s$ (Number of Fourier Series Pairs)

Model	p	R_F^2 (R^2)	SSE_F	AIC_F	$R_{F,pred}^2$ (R_{pred}^2)
Saturated:	17	0.6074	119.60	1064.59	0.5323
or $= 5$, pr $= 6$		0.7235			0.6698
or $= 5$, pr $= 5$	15	0.6062	119.95	1061.23	0.5420
		0.7230			0.6768
or $= 5$, pr $= 4$	13	0.6060	120.02	1057.35	0.5508
		0.7229			0.6830
or $= 5$, pr $= 3$	11	0.6022	121.20	1055.45	0.5555
		0.7211			0.6873
or $= 5$, pr $= 2$	9	0.5992	122.10	1053.05[a]	0.5609[a]
		0.7194			0.6915
or $= 4$, pr $= 2$	8	0.5902	124.83	1055.80	0.5548
		0.7100			0.6840
or $= 3$, pr $= 2$	7	0.5804	127.84	1058.92	0.5499
		0.6958			0.6736

[a] Best model.

Models simpler than the ones shown display residuals not diagnosed as AR(0) using *ar.yw( )* and are understood as having dropped signal back into the noise.

The usual graph for the model, separating the periodic and trend components, is displayed in Figure 16.14. As with the Mauna Lua CO_2 data, each figure is based on the assumption that the component being ignored is deterministic and the variation is all in the component being plotted. Although the units are not given for this data, it is clear that the seasonal variation is ± 1.5 units, while there has been a drop in the

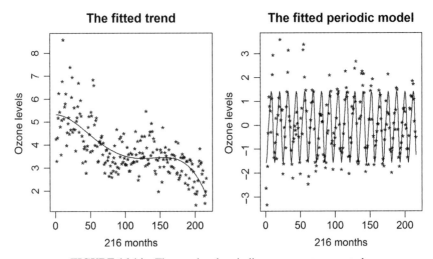

FIGURE 16.14   The trend and periodic components separated.

average from about 5.2 units to 1.8 units. Although the curve is quite complex, it is clear there has been a steady drop in average ozone level for the given time period (the curve is monotone decreasing).

16.5 PERIODIC MODELS: MONTHLY, WEEKLY, AND DAILY AVERAGES

Whether data is gathered monthly, weekly, or daily (or any other interval) has a huge impact on the type of model that results. Generally, sampling at different intervals will change the nature of the serial correlation, the less frequency the sampling, the less dependence between observations. Furthermore, many measurements are summed over intervals, so that variability is averaged out, when the time frame increases. Consider an example. To avoid some irregularities of real data (to be discussed shortly), years are constructed with 360 days and weeks with 6 days. Data like this has 12 months, 60 weeks, and 360 days.

```
#Simulating ten years of simple periodic data
n <- 3600
time <- c(1:3600)
error <- arima.sim(list(ar=c(0.6, -0.51, -0.28)), n = 3600)
y <- 15 + 3*cos(2*pi*time/360) + 4*sin(2*pi*time/360)
daily <- y + 3*error
# fitting a model to the data
col_1 <- cos(2*pi*time/360)
col_2 <- sin(2*pi*time/360)
fit_day <- lm(daily ~ col_1 + col_2)
plot(time, daily, pch = ".")
lines(time, fit_day$fitted)
title("The data measured daily")
```

For this data in Figure 16.15, the explained variation is $R^2 = 0.3127$, and the residuals are AR(3), based on $ar.yw()$.

Suppose these same measurements are accumulated for 6 days and reported "weekly." For this data, displayed in Figure 16.16, the explained variation is $R^2 = 0.89$, and the residuals are AR(2), based on $ar.yw()$.

Finally, the data could be measured monthly (every 30 days). For this data, displayed in Figure 16.17, the explained variation is $R^2 = 0.9792$, and the residuals are AR(1), based on $ar.yw()$. To reiterate, average over observations has reduced variance and producing observations spaced further apart has reduced the order of the serial correlation.

Of course, real data has further complications. There are leap years; months differ in length; and a year is not an integer number of days. For example, if data is measured daily for long periods of time, a more exact period would be 365.242 days per year. There is nothing wrong with a model with terms like $\cos(2\pi \cdot t/365.242)$ included

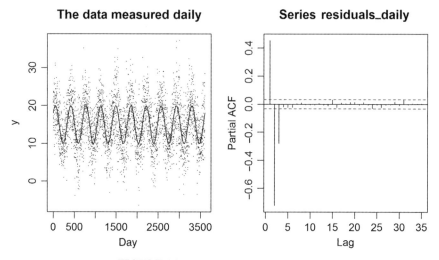

FIGURE 16.15   The data measured daily.

(the period would almost repeat every 365 days and a leap day every 4 years would get closer still). Data collected weekly also has the problem that there are not exactly 52 weeks in a year. Because $365.242/7 \approx 52.177$, an approach for weekly data would involve terms like $\cos(2\pi \cdot t/52.177)$. Part of the issue is that week 3 of 2013 and week 3 of 2014 are not at exactly the same time in the annual cycle. Using these apparently irregular cycles, and allowing for various corrections such as leap days, keeps the cycles close over long periods of time.

For monthly data, there are always 12 months in a year, but the number of days per month varies. Usually it makes more sense to compute averages rather than totals for any accumulated values collected over a month.

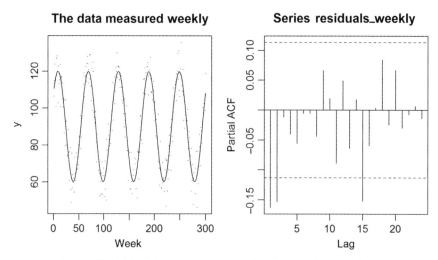

FIGURE 16.16   Measurements summed and reported every 6 days.

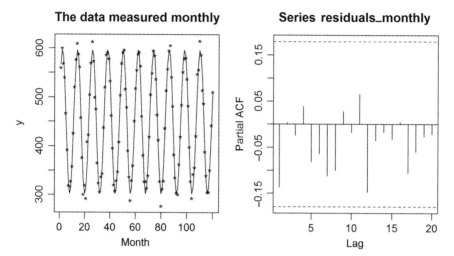

FIGURE 16.17   Measurements summed and reported every 30 days.

16.6 SUMMARY

In this chapter, a lot of different models have been fitted to the data, using basic regression ideas: periodic models, local smoothers, period + trend, and intervention models. What makes the modeling here different from a standard regression textbook is that a saturated model is used to determine residuals. In turn, the residuals are used to determine the order, m, of an AR(m) to filter the data. Model selection is then performed on the filtered data. Inference is also performed on the filtered data, although plotting the model with the original data and computing R^2 and R^2_{pred} can be informative.

EXERCISES

1. Let $b(B) = (1 - 0.6B)^2(1 + 0.4B) = 1 - 0.8B - 0.12B^2 + 0.144B^3$
    (i) Simulate errors for this model with $n = 5000$ and standard deviation 1.0.
    (ii) Plot *acf(error)* and *pacf(error)*.
    (iii) Use *ar.yw()* or *ar.mle()* to find the correct order, m, for an AR(m) model and the values $\hat{a}_1 \ldots \hat{a}_m$.
    (iv) Filter the errors.
    (v) Verify, using *ar.yw()* or *ar.mle()*, that AR(0) is the model selected for the filtered errors.
    (vi) Plot *acf(filtered error)* and *pacf(filtered error)*.

2. (There could be many reasonable approaches to this exercise. A good statistician will wrestle with this question all their life.) Consider a unique event such as the 2037 Super bowl of the 3011 Indianapolis 500. Presumably there is some chance that the Green Bay Packers will win the 2037 Super bowl. Given that the

event is unique, how can there be a probability that the Green Bay Packers will win? After all, probabilities are based on long-run averages. Although it could be argued there can be no probability, it is the opinion of the author that there is a probability and that this probability is less than 0.5 and greater than 1/5000. In fact, other than the trivial examples of flipping coins and rolling dice discussed in probability courses, what events are not unique?

3. Consider the AR(2) model $(1 - 0.7B)(1 + 0.1B)\varepsilon_n = w_n$:
   (i) Find a_1 and a_2.
   (ii) Simulate errors for this model with $n = 300$ and $\sigma = 1.5$.
   (iii) Construct a filtering matrix A using a_1 and a_2.
   (iv) Filter the errors and use *pacf()* and *ar.yw()* [or ar.mle()] to demonstrate that the filtering has been successful.

4. Consider the AR(3) model

$$(1 - 0.7B)(1 - [0.4 - 0.1i]B)(1 - [0.4 + 0.1i]B)\varepsilon_n = w_n:$$

   (i) Find a_1, a_2 and a_3.
   (ii) Simulate errors for this model with $n = 400$ and $\sigma = 2.5$.
   (iii) Construct a filtering matrix A using a_1, a_2, and a_3.
   (iv) Filter the errors and use *pacf()* and *ar.yw()* [or *ar.mle()*] to demonstrate the filtering has been successful.

5. Fit a periodic model to the Ozone, Arosa (or Ozone Azusa) data in The Exercises folder ("Ozone Arosa.txt" and "Ozone Azura.txt", Exercises folder or DataMarket – Time Series Data Library – Meteorology) using a Fourier series (paired trigonometric functions). Use model selection and filtering to find the best model.

6. Use a trend and periodic model to fit the monthly milk production data ("Milk production.txt", Exercises folder). Use model selection and filtering to find the best model. When are the low and high months for milk production? How much is milk production increasing per month?

7. Use a trend and periodic model to fit the monthly Boston armed robberies data ("Boston robberies.txt" in the Exercises folder or DataMarket–TSDL–Crime– Monthly Boston armed robberies January 1966–October 1975). Use model selection and filtering to find the best model. When are the low and high months for armed robberies? How much are armed robberies increasing per month?

8. Perform an intervention analysis for the monthly Minneapolis public drunkenness data ("Minn drunks.txt" from the Exercises folder or DataMarket–TSDLme– Monthly Minneapolis public drunkenness intakes January 1966–July1978). How much did drunkenness drop (include an interval), and when?

9. Perform an analysis of the passenger miles flown in the UK ("UK road deaths.txt" from the Exercises folder). Use model selection and filtering to find the best model. Are passenger miles increasing? If so, by how much?

10. For the models compared in Table 16.9 (LA ozone), the models were all nested. Use hypothesis testing to pick the best model.

11. (i) Simulate $n = 800$ observations with a simple periodic function, a period of 100, and errors of form AR(3) given in exercise 4, choose the simple periodic function so that initially $0.75 < R^2 < 0.95$.

  (ii) Fit a simple periodic function to the data and use the residuals to assess the order AR(m) and compute R^2 for the model fit.

 (iii) Average over 5-day periods, fit a simple periodic function to the data and use the residuals to assess the order AR(m) and compute R^2 for the model fit.

 (iv) Average over 10-day periods, fit a simple periodic function to the data and use the residuals to assess the order AR(m) and compute R^2 for the model fit.

  (v) Average over 20-day periods, fit a simple periodic function to the data and use the residuals to assess the order AR(m) and compute R^2 for the model fit.

 (vi) Summarize how the AR(m) order and R^2 change as greater averaging is introduced.

12. Produce prediction and prediction intervals for AR(2) errors $\varepsilon_{j+1}, \varepsilon_{j+1}$, and ε_{j+1}, similar to Section 16.2.5.

PART IV

SOME DETAILED AND COMPLETE EXAMPLES

17

WOLF'S SUNSPOT NUMBER DATA

17.1 BACKGROUND

From Wikipedia ("Wolf number" and "Solar Storm of 1859"), although this is all well known.

> The Wolf number (also known as the International sunspot number, relative sunspot number, or Zürich number) is a quantity that measures the number of sunspots and groups of sunspots present on the surface of the sun.
>
> The idea of computing sunspot numbers was originated by Rudolf Wolf in 1848 in Zurich, Switzerland and, thus, the procedure he initiated bears his name (or place). The combination of sunspots and their grouping is used because it compensates for variations in observing small sunspots.
>
> This number has been collected and tabulated by researchers for over 150 years. They have found that sunspot activity is cyclical and reaches its maximum around every 9.5 to 11 years (note: Using data from SIDC for the last 300 years and running a FFT function on the data gives an average maximum at 10.4883 years/cycle). This cycle was first noted by Heinrich Schwabe in 1843.
>
> The relative sunspot number R is computed using the formula (collected as a daily index of sunspot activity): $R = k(10g + s)$ where s is the number of individual spots, g is the number of sunspot groups, and k is a factor that varies with location and instrumentation (also known as the *observatory factor* or the *personal reduction coefficient K*).

Sunspot activity, in general, is associated with specific events known as coronal mass ejections and solar flares (the degree to which these similar events are related

Basic Data Analysis for Time Series with R, First Edition. DeWayne R. Derryberry.
© 2014 John Wiley & Sons, Inc. Published 2014 by John Wiley & Sons, Inc.

is unclear). These events can be quite dramatic even as experienced 93 million miles away on earth. A solar flare in 1859, the "Carrington event," reported by British astronomer Richard Carrington, produced auroras observable in Hawaii and Cuba and set telegraph lines on fire. Periods of great sunspot activity can disrupt electrical services and short wave radio communications, accelerate the orbital decay of satellites, and could do harm to astronauts.

17.2   UNKNOWN PERIOD ⇒ NONLINEAR MODEL

For the sunspot data (Figure 3.2, from year 1700 to 1988), the exact period is unknown. While others have found the period to be 9.5 to 11 years (as mentioned in the Wikipedia quote above), a good beginning requires verifying this, rather than accept it as fact.

To fit an exact period, a model is required that allows the period to vary as a free parameter. A nonlinear least squares problem can be formulated:

$$\min \sum \left(y_j - \mu - B \cdot \cos[2\pi j/k] - C \cdot \sin[2\pi j/k] \right)^2,$$

where μ, B, C, and k are all unknown (Previously k was treated as known).

This problem is not a linear least squares problem because the partial derivatives, $\frac{\partial}{\partial \mu}, \frac{\partial}{\partial B}, \frac{\partial}{\partial C}$, and $\frac{\partial}{\partial k}$, when set equal to zero, although they do form a system of four equations in four unknowns, do not result in a system of *linear* equations, and so cannot be solved using linear algebra.

17.3   THE FUNCTION *nls()* IN R

There is an R function, *nls()*, designed to solve more general least squares problems using general optimization methods. Optimization techniques do not always converge to a good solution, and a number of issues are important when dealing with such methods [Appendix C, written from the point of view of a user, discusses issues related to nonlinear optimization in the context of *nls()*]. One issue, to be dealt with immediately, is that these methods, unlike ordinary least squares, require starting guesses for the unknown values. Sometimes it matters little what the starting values are, other times the function will fail to converge due to poor initial guesses.

To introduce *nls()*, a problem is solved in two different ways, with *lm()* and with *nls()*.

Consider the model $y = ae^{bx} + \varepsilon$. Assuming the errors are white noise, this problem could be formulated as $\min \sum \left(y_j - ae^{bx_j} \right)^2$. It is easy to verify that, when set equal to zero, $\frac{\partial}{\partial a}$ and $\frac{\partial}{\partial b}$ do not form two linear equations in two unknowns. In order to avoid the nonlinear nature of this problem, it is quite common to transform the data to form a linearized version of this problem. Reformulated, this is $\log_e(y) = \log_e(ae^{bx} + \varepsilon) \approx \log_e(a) + bx + \varepsilon^*$. This problem can be solved directly using *lm()*. It is hoped that the new errors, ε^*, are also approximately white noise, and often this transformation brings right skewed errors closer to white noise.

After simulating some data, both methods will be used to solve for *a* and *b*.

```
# simulating data based on an exponential model
a <- 2
b <- 0.25
x <- c(-50:50)/20
y <- a*exp(b*x)
n <- length(x)
error <- rnorm(n,0,0.2)
y_err <- y + error
plot(x,y_err)
lines(x,y)
title("A nonlinear model")
plot(x,log(y_err))
lines(x,log(y))
title("The model linearized")
```

It is easy to see why linearization is attractive (Figure 17.1). Although some injustice has been done to the errors, the function *lm()* is far more well behaved than *nls()*, which may require some sophisticated efforts to produce convergence.

fit <- lm(log(y_err) ~ x) produces the output:

	Estimate	Std. Error	t value	Pr(>|t|)
(Intercept)	0.691546	0.010805	64.00	<2e-16
x	0.254504	0.007412	34.34	<2e-16

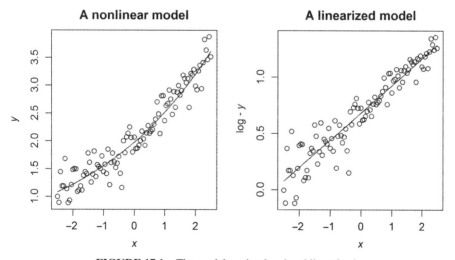

FIGURE 17.1   The model as simulated and linearized.

Recall, $\hat{a} = \exp(0.691546) = 1.9968$ and $\hat{b} = 0.2545$. The estimates are quite good.

The *nls()* approach is specified as follows [notice that the solutions from *lm()* provide initial guesses for *nls()*]: fit_nls <- nls(y_err ~ a*exp(b*x), start=list(a=1.9968, b = 0.254504)).

The two main differences between *lm()* and *nls()* are: (i) in *lm()*, the functional form of the model is presumed to follow a certain linear pattern; in *nls()*, the functional form of the model must be specified and (ii) initial guesses must be given for the unknowns. The result is summary(fit_nls):

	Estimate	Std. Error	t value	Pr(>|t|)
a	2.000653	0.021020	95.18	<2e-16
b	0.259352	0.006589	39.36	<2e-16

Often, the starting values for the function are found using a less precise method, known to produce reasonable guesses. Obviously, the linearized model should produce excellent initial guesses. In this case, the *lm()* solutions are so good; *nls()* produces only marginal improvement.

17.4   DETERMINING THE PERIOD

In the case of an unknown period, a simple estimate is quite easy. If there are clear distinct peaks, a rough estimate of the period is $k \approx n/peaks$. For the sunspot data, there are 27 distinct peaks and 289 observations, so $k \approx 289/27 = 10.7$. This will be the initial guess for *nls()* unless other graphs suggest this is a problem.

The *acf()* plot (positive peaks around 10 and 20, negative peaks around 5 and 15) and the periodogram in Figure 17.2 both consistent with an estimated period of about 10.7 years.

Given a reasonable guess about k, it is then possible to use the simple periodic model to get guesses for B, C, and μ. These were found to be (verified in an exercise) $\hat{\mu} = 6.3434$, $\hat{B} = -0.1919$, and $\hat{C} = 0.6404$. A good guess for the period has been found, and good guesses for the other three unknowns, given the period, have been found. With reasonable starting values, it is possible to enlist *nls()*. The *nls()* fit is then found with the following code: fit_nls <- nls(sqrt(y_sunspots) ~ mu + B * cos(2*pi*time/k) + C * sin(2*pi*time/k), start = list(mu = 6.3434, B = -0.1919, C = 0.6404, k = 10.7))

	Estimate	Std. Error	t value	Pr(>|t|)
mu	6.35732	0.15866	40.069	<2e-16
B	-0.57681	0.42755	-1.349	0.178
C	-1.43947	0.26786	-5.374	1.6e-07

| k | 10.50626 | 0.03046 | 344.879 | < 2e-16. So, the estimate of the period is $\hat{k} = 10.50626$.

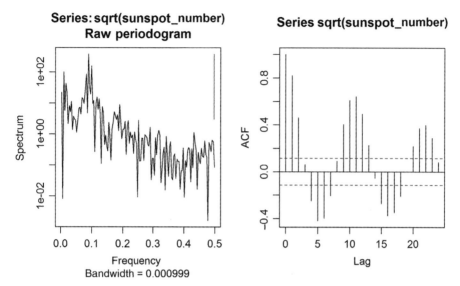

FIGURE 17.2   The sunspot data, with the square root transformation, periodogram, and *acf( )* plot.

To find fitted values for the model (in order to plot the fit), the location of the estimated coefficients must be found names(summary(fit_nls)):

```
[1] "formula"     "residuals"    "sigma"       "df"          "cov.unscaled"
[6] "call"        "convInfo"     "control"     "na.action"   "coefficients"
[11] "parameters"
```

Recall that for any statistical model: fitted value = observation – residual. So the R code: fitted_model <- sqrt(sunspots) - summary(fit_nls)$residual produces fitted values.

17.5   INSTABILITY IN THE MEAN, AMPLITUDE, AND PERIOD

This fitted model will not, however, provide a good fit to the data (see Figure 17.3). The mean, period, and amplitude all vary for this data. Using a method of moving averages, or moving windows, the variation in these values over time is explored (the method used here is very different from complex demodulation, but the results are quite similar, see Bloomfield, 2000, Chapter 7).

Continuing with Figure 17.3, it is obvious from looking carefully at the data that the period is highly variable. Notice that from 1760 to almost 1788, very few years (8–10) made up a full period, In fact, an up movement from a minimum often took only 3–4 years, and a down movement took 5–6 years. However, from about 1789 to 1800, in an 11-year period, only one down movement to a minimum occurred (right panel of Figure 17.3).

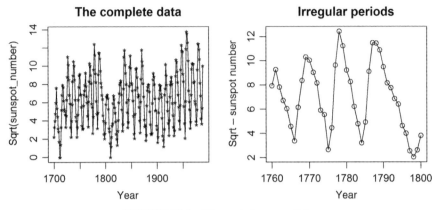

FIGURE 17.3   The periods vary wildly.

If the model does not vary over time, different subsets of the data should produce similar estimated models for the data. On the other hand, if the model does vary over time, local subsets of the data will produce very different models. One way to test this claim is to examine the data locally, fit the simple periodic model to subsets of the data, and then compare the model with one fixed set of parameters to the model with parameters that vary based on local data.

A window is a period of years over which observations will be included in each local simple periodic model. By moving the window one observation at a time, a local model can be built around each point. If the model does change slowly, seeming to vary only slightly from observation to observation, it might be that one model fits all of the data. However, if the model changes rapidly, the underlying model may be varying over time. Using model selection, a relatively definitive answer may be possible.

For example, if the window is 31 years (about three periods), then each observation in the middle of the data will have a local model built using the observation, the 15 previous years, and the 15 following years. The first and last 15 years of data do not have enough observations to build a window of size 31. A number of possible approaches to the endpoints are possible. The convention here will be to fit the endpoints based on model determined by the closest full window. In other words, the first 15 years will have the same model as observation 16, and the last 15 years (years 275 through 289) will have the same model as observation 274.

As a preliminary analysis, to develop a feel for such a model, the period is set at 10.50625 and windows of 31 observations are used to build a model. Taking each point as the center of a new window, a plot of how μ, M, and ϕ vary can be produced (Figure 17.4, the right panel). Compared to the left panel of this same figure, which is the fit to the data for a simple periodic function with fixed parameters for all of the data, the model with varying parameters seems to fit the data much better. However, it is not clear if this is really the case until some account is taken for the fact that the models have very different levels of complexity (it should be noted that no filtering has yet been employed, so this is all exploratory).

The following code is used to create the graph on the right panel in Figure 17.4. The following code is quite complicated. The main "for" loop fits the model based

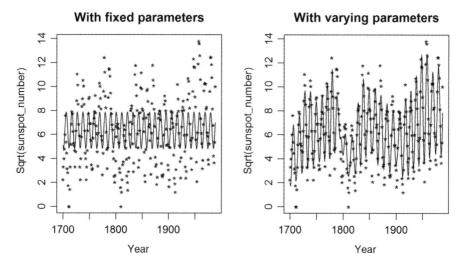

FIGURE 17.4 The models for the sunspot data with fixed and variable parameters.

on windows of size 31, to observations 16 through 274. The code is written so that the window size (*w*) can be varied. The period is fixed at 10.50626.

```
# the values for the parameters will be stored for each window. A total of 259 interior
#points will be fitted. The first fit will be based on observations 1 to 31 and the last fit
#will be  based on observations 259 to 289. The code can be customized for any
window # size by changing w (for model selection).
w <- 19                 # the size of the window for each fit
pad <- (w-1)/2          # the number of observations not fitted on each end
                        # the model will be extended, by extrapolation, to these

                        # observations at the end of the code.

iters <- 289-2*pad      # the number of times to go through the for loop
store_mu <- rep(NA,iters)      # store estimates
store_B <-rep(NA,iters)
store_phi <- rep(NA,iters)
store_C <- rep(NA,iters)
store_M <- rep(NA, iters)
time <- c(1:289)
z <- sqrt(y_sunspots)
# coefficients from each  lm() fit are stored in a storage vector
for(j in 1:iters)
{y <- z[j:(j+w-1)]
tmp <- time[j:(j+w-1)]
col_c <- cos(2*pi*tmp/10.50626)
col_s <- sin(2*pi*tmp/10.50626)
```

```
fit_tmp <- lm(y ~col_c+col_s)
store_mu[j] <- fit_tmp$coeff[1]
store_B[j] <- fit_tmp$coeff[2]
store_C[j] <- fit_tmp$coeff[3]}


#
# obtain estimates of the amplitude (M), and phi
#
store_M <- sqrt(store_B^2 + store_C^2)
store_phi <- atan2(-store_C,store_B)
store_phi <- store_phi/(2*pi)
# Extend, by extrapolation the model at the endpoints to fit all of the data.
# There are "pad" = (w-1)/2 values at each end that do not have a fitted model.
# For example, the observations 1...pad will be assigned the model for
# observation pad+1
store_M <- c(rep(store_M[1],pad),store_M, rep(store_M[iters],pad))
store_phi <- c(rep(store_phi[1],pad),store_phi, rep(store_phi[iters],pad))
store_mu <- c(rep(store_mu[1],pad),store_mu, rep(store_mu[iters],pad))
store_B <- c(rep(store_B[1],pad),store_B, rep(store_B[iters],pad))
store_C <- c(rep(store_C[1],pad),store_C, rep(store_C[iters],pad))


# It is possible to use this to fit a model to the data be combining the time varying    #
components as a simple trigonometric function
fitted_model    <-    store_mu    +    store_B*cos(2*pi*time/10.50626)    +
store_C*sin(2*pi*time/10.50626)
resid <- sqrt(y_sunspots) - fitted_model  # the residuals
```

It should be noted that the R function *atan2(-B,A)* replaces the elaborate branch-wise formula of Section 10.1 for computing $\hat{\phi}$, given $\hat{B}$ and $\hat{C}$. As always, *help(atan2)* is useful.

How are the individual parameter estimates changing? Figure 17.5 displays the variation in each of the periodic function parameters over the 289-year period.

Figure 17.5 suggests that something quite unusual may have happened around 1810, with a less severe, but also unusual pattern just prior to 1900. The noise has quite a complex AR(*m*) structure as well, based on the *pacf()* plot.

Interestingly, *ar.yw()* found the residuals (residuals <- sqrt(y_sunspots) - fitted_model) from the variable parameter-fitted model to be of order 20!

17.6   DATA SPLITTING FOR PREDICTION

17.6.1   The Approach

For the sunspot data, an interesting question is, which window size is best for making future predictions. Perhaps the routine of the previous sections leading to a model

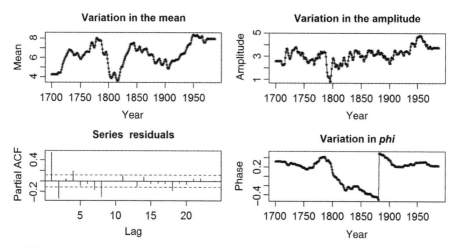

FIGURE 17.5 Variations in the phase, amplitude, and mean over time. Also the *pacf()* plot of the residuals from the variable parameter model.

with the best $R^2_{F,\text{pred}}$ might make sense. However, with a constantly changing local model, it might make more sense to evaluate how the model fits new prediction moving forward through the known data. In other words, treat the first half of the data as a training set and the second half as a validation set. This takes the notion of extrapolation head on. This assesses how well a fit would have been found if predictions were made, 1 year ahead of time, for the last 144 years of the data.

For any given window size w, the approach works as follows:

(i) Using the first half of the data, fit a moving local period function to the data.

(ii) Use the residuals to diagnose the order, m, for an AR(m) model using ar.yw().

(iii) Extrapolate the model one (or more) step(s) ahead as the next prediction and use the last m residuals for an AR correction to this prediction.

(iv) Keep moving the model ahead one step at a time and repeat (iii) for each new observation in the validation set.

A one-step ahead prediction at the point j, given any window up to and including $j-1$ is of the form:

$$\hat{y}_{\text{AR},j} = \mu_{j-1} + B_{j-1}\cos(2\pi j/k) + C_{j-1}\sin(2\pi j/k) + \hat{a}_1(y_{j-1} - \hat{y}_{j-1}) \cdots \\ + \hat{a}_m(y_{j-m-1} - \hat{y}_{j-m-1}).$$

The value $R^2_{\text{AR}} = 1 - \sum(y_j - \hat{y}_{\text{AR},j})^2 / \sum(y_j - \bar{y})^2$, where the y's are from the validation set only, is used to pick the best model. Notice that no filtering is done here. Instead, the model selection is based on a direct comparison of predictions versus observed values; the AR(m) structure of the residual is, however, important, as corrections are made to the model prediction based on the last m residuals using $\hat{a}_1 \ldots \hat{a}_m$ from ar.yw(). With minor modifications, a two- or three-step ahead method can be used.

17.6.2 Step 1—Fitting One Step Ahead

Code, in R that fits a model one step ahead, in other words, finds $\hat{y}_{AR,j} = \mu_{j-1} + B_{j-1}\cos(2\pi j/k) + C_{j-1}\sin(2\pi j/k)$, is as follows:

```
n <- 289
w <- 43     # the window size for all fits
y <- sqrt(y_sunspots)
y_fit <- rep(NA,n)
k <- 10.50626
# fit to the first w values
tm <- c(1:w)
col_c <- cos(2*pi*tm/k)
col_s <- sin(2*pi*tm/k)
y_first <- lm(y[tm] ~ col_c+col_s)
y_fit[1:w] <- y_first$fit
# the fit to the rest of the data.In the for loop "j" is the end of the current window, and
# the "j"th observations is fitted.
for(j in w:(n-1))
{ tme <-c((j-w+1):j)
col_c <- cos(2*pi*tme/k)
col_s <- sin(2*pi*tme/k)
fit_tmp <- lm(y[tme] ~ col_c+col_s)
y_fit[j+1]              <-         fit_tmp$coeff[1]+fit_tmp$coeff[2]*cos(2*pi*(j+1)/k)
+fit_tmp$coeff[3]*sin(2*pi*(j+1)/k)}
```

17.6.3 The AR Correction

The next step is to find $\hat{a}_1(y_{j-1} - \hat{y}_{j-1}) \cdots + \hat{a}_m(y_{j-m-1} - \hat{y}_{j-m-1})$. This involves assessing the order of the residuals, storing the values $\hat{a}_1 \ldots \hat{a}_m$, and combining them with the correct residuals.

```
residuals <- y - y_fit
z <- ar.yw(residuals[1:145])   # using the training set for diagnosis of the residuals
# making AR predictions
m <- z$order
ar_corr <- rep(0, n)                 # the corrections to be added to the predictions
resid_tmp <- rep(NA,m)
ar <- z$ar
#combining the residuals with the
for(j in m:(n-1))
{ resid_tmp <- residuals[(j-m+1):j]
sum <- 0
for(k in 1:m)
{ sum <- ar[k]*resid_tmp[m+1-k]+sum}
ar_corr[j+1] <- sum}
```

17.6.4   Putting it All Together

The corrections and predictions are combined to produce the usual AR-adjusted predictions.

```
y_AR <- y_fit+ar_corr
residuals_AR <- y - y_AR
```

How good are the models? Plot the predictions and corrected predictions on the validation part of the data (observations 146 to the end).

```
validation_period <- c(146:n)
plot(validation_period,y[validation_period], ylab="sqrt(sunspot numbers)")
lines(validation_period,y_fit[validation_period])
lines(validation_period, y_AR[validation_period],lty = 2)
title("Predictions(solid) and corrected predictions (dashed)")
```

When the window size is large, the initial predictions are not very accurate, but the AR corrections lead to substantial improvement, as can be seen in Figure 17.6.

17.6.5   Model Selection

Model selection is quite straightforward. Different window sizes are considered, and the one with the best value for R^2_{AR} (where the fit it to the AR-corrected predictions) is best.

The results displayed in Table 17.1 reflect the power of making AR-based corrections, at least in this case. If the AR corrections are used, one can be quite casual

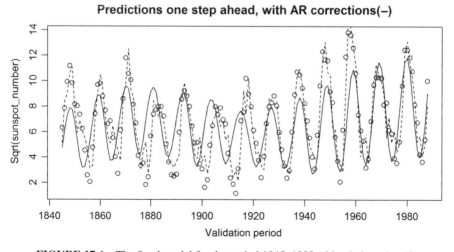

FIGURE 17.6   The fitted model for the period 1845–1988 with window size 43.

TABLE 17.1   Cross Validation, Looking One Step Ahead, for the Sunspot Data ($n = 145$)

w	m	R^2	R^2_{AR}
3	7	0.4991	0.8440
5	3	0.6915	0.8229
7	2	0.7026	0.8505
9	2	0.7332	0.8349
11	2	0.7524	0.8584[a]
13	2	0.7328	0.8583
15	2	0.7288	0.8576
17	2	0.7208	0.8560
19	2	0.7046	0.8495
21	2	0.6603	0.8476
23	2	0.6391	0.8486
33	7	0.5409	0.8449
43	7	0.4297	0.8426
53	7	0.3347	0.8435

[a]Best model.
Source: 2010 query of the National Cancer Institute – Surveillance, Epidemiology and End Results database, used with permission.

about the choice of window size and get reasonable predictions. Given that the AR model is built in the training set, and no updating of the AR model is done, going forward in time, this seems amazing. Notice that the models $w = 7$ to $w = 23$ only require AR(2) corrections. The best models seem to be about one to two periods (10 to 20 years) in length.

The code for computing the R^2 values was

```
SST <- var(y[validation_period])*(144-1)
SSE <- sum(residuals[validation_period]^2)
SSE_AR <- sum(residuals_AR[validation_period]^2)
R2 <- 1-SSE/SST
R2_AR <- 1-SSE_AR/SST
```

17.6.6   Predictions Two Steps Ahead

The formula makes everything clear:

$$\hat{y}_{AR,j+1} = \mu_{j-1} + B_{j-1}\cos(2\pi[j+1]/k) + C_{j-1}\sin(2\pi[j+1]/k)$$
$$+ \hat{a}_1\hat{r}_j + \hat{a}_2(y_{j-1} - \hat{y}_{j-1}) \cdots + \hat{a}_m(y_{j-m} - \hat{y}_{j-m}),$$

where $\hat{r}_j$ must be estimated using the previous m residuals.

In other words, there is actually no information about the last residuals before the predictions (because it has not yet been observed) and the model must be extrapolated two steps into the future. The required changes, to the R code, are minor.

TABLE 17.2 **Cross Validation, Looking Two Steps Ahead, for the Sunspot Data ($n = 145$)**

w	m	R^2	R^2_{AR}
3	8	0.000	0.000
5	18	0.0463	0.6056
7	11	0.4024	0.5802
9	3	0.6349	0.6455
11	3	0.6696	0.7021[a]
13	3	0.6635	0.6967
15	3	0.6498	0.6844
17	3	0.6493	0.6801
19	11	0.6132	0.5706
21	7	0.5869	0.6378
23	2	0.5809	0.6401
33	7	0.4796	0.6289
43	7	0.3551	0.6159
53	7	0.2524	0.6093

[a]Best model.
Source: 2010 query of the National Cancer Institute – Surveillance, Epidemiology and End Results database, used with permission.

Overall, the results are as expected (Table 17.2). Making predictions 2 years ahead of time degrades the predictions quite a bit, but AR corrections still produce improvements (Figure 17.7) and are more robust to model selections choices. In this case, as before, the window size 11 seems to be the best. For two-step ahead prediction, window size 19 produced some quite anomalous results, with a very large order of AR(11), and poorer residuals with correction than without.

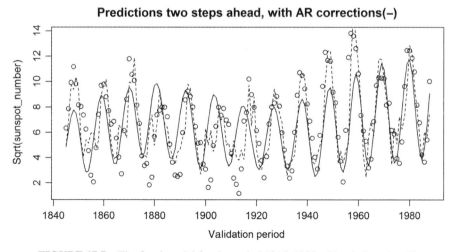

Predictions two steps ahead, with AR corrections(–)

FIGURE 17.7 The fitted model for the period 1845–1988 with window size 43.

17.7   SUMMARY

The sunspot data can be fitted using a simple periodic model that changes over time. Using data-splitting ideas in a new way, it is possible to compute values of R^2_{AR} that represent the predictive accuracy of the model if current data were used to make predictions 1–2 years (or more) in advance of the current data. For predictions 1 year ahead, $R^2_{AR} \approx 0.85$ is realistic and for predictions 2 years ahead, $R^2_{AR} \approx 0.70$ is realistic. These predictions are not very reliable unless AR(m) corrections are made for each prediction, although using $m = 2, 3$ is often sufficient.

EXERCISES

1. Verify the estimates of μ, B, and C for the simple periodic model when $k = 10.7$.

2. Consider the model $y = 3 + 0.5x + 0.2x^2 + \varepsilon$, for $-2 < x < 2$.
   (i) Simulate $n = 50$ observations from this model with a definite signal and some definite noise, perhaps $0.7 < R^2 < 0.9$.
   (ii) Fit a model using $lm()$.
   (iii) Fit a model using $nls()$.
   (iv) Compare the results.

3. This exercise uses the file "LYNX.txt" from the Exercises folder.
   (i) Assess whether a transformation is required for the Lynx data. If a transformation is appropriate, use it to complete all steps of the exercise.
   (ii) Estimate the period of the lynx data and use $nls()$ to refine this estimate.
   (iii) Fit a model with fixed period to the data and graph this model versus the data. How good does this model look (compute R^2)?
   (iv) Fit a model with about three complete periods per window, and a fixed period, but allowing B, C, and μ to vary. Plot the resulting model versus the data. How much better is this model (compute R^2)? Make a graph like Figure 17.5 showing how much the mean, phase, and amplitude vary.
   (v) Use model selection based on the one-step ahead model to find the best window size. How similar to or different from the sunspot data is this.

4. Modify the R code from the book to create a two-step ahead method and apply this to the Lynx data.

5. Repeat the steps form the Lynx exercise with the file "Zuerich sunpots.txt" from the Exercises folder (DataMarket–Time Series Data Library–Physics–Zuerich monthly sunspot numbers 1749–1983). Think about how monthly data would differ from annual data – remember, windows should be thought of as related to period (the use of common sense is permitted in statistics).

6. In the one-step ahead model selection table (Table 17.1), the models with very small or very large windows have more complex structure in the residuals. Is this

evidence that these models are not as appropriate as the models $w = 7$ to $w = 23$? Explain.

7. (An open-ended problem). The one- and two-step ahead methods did not use the data as efficiently as possible, because the AR(m) model was built on the training set and never updated, as the model moved forward through time. Assess whether the structure of the residuals changes over time, and revise the one-step ahead code to better model the noise. Does this improve the model fit? Explain.

8. (A very open-ended problem). Perhaps more a project than an exercise: simulate some periodic data with a varying mean, period, phase, and amplitude, but with a moderate amount of random ARMA(2,2) noise. Use the model selection methods of the book and assess how well the method works for prediction.

18

AN ANALYSIS OF SOME PROSTATE AND BREAST CANCER DATA

18.1 BACKGROUND

It has been observed that both breast cancer and prostate cancer are rarer near the equator and more prevalent toward the poles. This is a worldwide phenomenon. Some of the research done by Dr. St Hilaire, Rakesh Mandal, and others highlights this pattern and gives a plausible explanation (Mandal et al., 2009; St-Hilaire, 2010, 2011).

These biologists suspect that pollutants combined with meteorological conditions cause areas far from the tropics to collect pollutants. The nature of prostate cancer and breast cancer are such that both could be caused by certain pollutants known as endocrine disruptors.

Some of the work involves:

(i) Showing that male prostate cancer and female breast cancer have an uncanny association (and part of this is showing cancers, in general, have little association).

(ii) Showing ER+ (estrogen response positive) and ER− (estrogen response negative) breast cancers have different prevalence patterns and that ER+ breast cancer seems to fit the "pollution-caused" model, while ER− breast cancer does not. More specifically, endocrine disruptors (and pollution in general) and ER+ breast cancer are both on the rise, while ER− breast cancer seems to be level over time.

Basic Data Analysis for Time Series with R, First Edition. DeWayne R. Derryberry.
© 2014 John Wiley & Sons, Inc. Published 2014 by John Wiley & Sons, Inc.

TABLE 18.1   Format of the Data from ER+, ER− Data Set

County	State	Year	ER+ rate	ER− rate
County A	A	1992	88.5	27.4
County A	A	1993	86.4	28.2
...				
County Z	H	1992	68.7	23.1

Source: Data were obtained from the National Cancer Institute-Surveillance, Epidemiology and End Results (SEER).

Each of these data sets indicates a way in which small sample sizes can be handled in the context of time series. Generally, time series analysis is based on large data sets, but reality does not always furnish large data sets. Nevertheless, the serial correlation in the data cannot be avoided or ignored. In the first case, each of several time series is reduced to a single number, and these different time series become a new data set without serial correlation.

18.2   THE FIRST DATA SET

The first data set ("ER+ ER− no names.txt" from the SSH folder. This data is not in DataMarket) produces an analysis that was never published, but the results highlight important ideas both for the research and for analysis of time series data.

Table 18.1 reflects the general structure of the original data, however, due to confidentiality issues, specific counties are not identified. The data includes breast cancer rates (estrogen response positive and negative) for 216 counties, in eight states, for the period 1992–2001 (10 years). The value "Rate" is an age-adjusted number of incidents per 100,000. The age adjustment is an epidemiological method that adjusts for the confounding variable age; however, the age-adjusted rates are close to an unadjusted number that would just be: rate = total incidents/total population.

The number of counties from each state is given in Table 18.2. The states were California, Connecticut, Georgia, Iowa, Michigan, New Mexico, Utah, and Washington. As before, for confidentiality reasons, the number of counties associated with each state is disguised.

Two things should be noted. There are a lot of data, but there is not much time series data. After all, with only a 10-year period, it is hard to model the proper AR(m) model, and filtering would wipe out some of what is already very scarce data.

TABLE 18.2   The States Included and the Number of Counties from Each State Reporting

State	A	B	C	D	E	F	G	H
Counties	99	33	29	17	14	12	8	4

However, the main question is the contrast between ER+ and ER− breast cancer rates over time. Often, even when each time series is reduced to a single number, there are still a lot of data for the purposes for which the data is intended.

In fact, even if each group of 10 numbers (the annual ER+ and ER− rate for each state and county for the period 1992–2001) is reduced to a single number, there are still 216 counties (from eight states) in which a paired comparison could be made within each county. In the non-time series world, this is a large data set, if it makes sense to treat individual counties as independent observations (it is always about independence).

In fact, trends within each time series can just be reduced to a single number, the slope of a regression line. When producing a number like this, which is just a point estimate, it is not problematic that the observations are serially correlated. The slope is a good estimate of what is happening within each time series; it is just that the standard errors for the slope are incorrect. But the standard errors are irrelevant to this analysis.

Therefore, the data set suggested in Table 18.1 can be reduced to 216 paired ER+ and ER− slopes. This produces a rather straightforward paired analysis.

The data ("ER+ ER− no names.txt") is read using the *read.table()* command. The columns are names(file name): "ER_plus" "ER_minus" "State".

For a full analysis, three interconnected questions need to be answered:

- Are ER− rates flat over the time period?
- Are ER+ rates increasing over the time period?
- Is the difference (ER+ rate − ER− rate) generally positive?

In each of these cases, confidence intervals, p-values, and histograms of the data are all useful.

Based on Figure 18.1 and a t-test, there is no evidence of an increase in ER− cancer rates (two-sided p-value = 0.362) for the period 1992–2001. The average rate of change in rates is estimated to be between −0.178 and 0.485. There is one potential outlier that, if excluded, or if a nonparametric analysis were performed, would almost certainly cause the p-value to be even larger. *It is plausible, based on this evidence that ER− cancer rates are not changing over time, for this time period.*

Based on Figure 18.2 and the t-test, there is convincing evidence (one-sided p-value less than 10^{-10}) that ER+ cancer rates are increasing, on average, over the period in question. The average increase is estimated to be between 1.64 and 2.62 incidents per 100,000 per year. *It is reasonable to conclude that ER+ rates are increasing over this particular time period, and that the increase is large.*

Based on Figure 18.3 and a paired t-test, there is convincing evidence (one-sided p-value less than 10^{-10}) that the difference in the change, over time, in ER+ minus ER− cancer rates (a difference between two slopes) within counties are, on average, positive. This difference is estimated to be between 1.41 and 2.54 incidents per 100,000 per year. *It is reasonable to conclude that, within counties, ER+ rates*

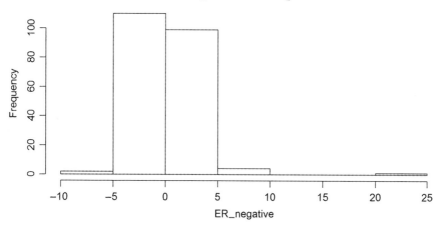

FIGURE 18.1   The slopes (change in cancer rate) for the period 1992–2001.

rise faster than, or diminish more slowly than, comparable ER– rates. Again, these numbers are practically, as well as statistically, significant.

A nonparametric analysis responding to the same three questions is left as an exercise.

The main point is as follows. It is often the case that multiple time series data are reduced to more traditional analysis by converting each time series into a single number and analyzing the data in this manner. In this data, even when the time series was reduced to a single number, there were still 216 pairs of numbers to compare, and the questions of interest could be addressed without further concern about issues of serial correlation. This was especially attractive in this case because the time series were too short to filter.

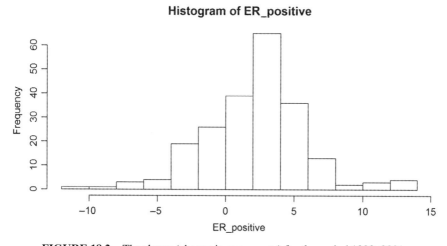

FIGURE 18.2   The slopes (change in cancer rate) for the period 1992–2001.

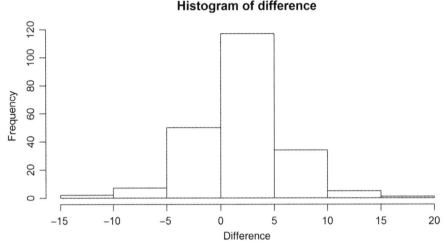

FIGURE 18.3   The difference in ER+ and ER− slope, county by county.

18.3   THE SECOND DATA SET

18.3.1   Background and Questions

The second data set involves cancer rates from the period 1975 to 2005 and is of the format shown in Table 18.3. This data set is in the SSH folder as "types of cancer 1975 2005.txt." This data is not available in DataMarket.

The rates (per 100,000) are for four types of female cancer (colon and rectum, lung and bronchus, ovary, and breast) and four comparable types of male cancer (colon and rectum, lung and bronchus, testis, and prostate). The columns are: names(filename)

[1] "Year" "Colon_f" "Lung_f" "Ovary" "Breast" "Colon_m" "Lung_m"
[8] "Prostate" "Testis".

The purpose of the research is to determine if there is an unusually strong positive association between breast cancer in females and prostate cancer in males. The

TABLE 18.3   The Structure of the Data for the Second Data Set

Year	Colon_f	Lung_f	Ovary	Breast	Colon_m	Lung_m	Prostate	Testis
1975	53.6437	24.524	16.3206	105.068	68.3998	89.5624	94.0202	3.7254
1976	54.1351	27.3104	15.8535	101.9126	71.7387	93.8146	97.9086	3.4415
1977	54.9506	28.3179	15.6194	100.8128	73.4974	95.5171	100.4533	4.3113
1978	55.2492	29.6523	15.3777	100.5789	72.3547	96.6654	99.3958	3.5846

Source: Data were obtained from the National Cancer Institute-Surveillance, Epidemiology and End Results (SEER).

belief is actually that there is an association between ER+ breast cancer and prostate cancer, as these are possibly linked by the previously discussed notion on endocrine disruption. However, it was not possible, in this data set, to separate breast cancer into the two types. If there is a link between ER+ breast cancer and prostate cancer, the inclusion of ER− breast cancer will dilute the link, but it should still be detected, especially because ER+ breast cancer is more prevalent than ER− breast cancer. This data is for the entire United States over this time period.

The argument is in two parts: (i) most cancers are not strongly positively associated, even within a sex, and (ii) female breast cancer and male prostate cancer are, despite this, strongly positively associated.

18.3.2   Outline of the Statistical Analysis

The analysis involves the following regressions:
How closely related are women's cancers? Perform three regressions:

log(breast cancer) ∼ log(colon and rectal)
log(breast cancer) ∼ log(lung and bronchus)
log(breast cancer) ∼ log(ovary)

How closely related are men's cancers? Perform three regressions:

log(prostate cancer) ∼ log(colon and rectal)
log(prostate cancer) ∼ log(lung and bronchus)
log(prostate cancer) ∼ log(testis)

How closely related are two cancers of interest? Perform one regression:

log(prostate cancer) ∼ log(breast cancer)

The logarithmic scale is appropriate because the relationship is probably proportional (a percentage increase in one cancer would be associated with a, perhaps different percentage increase in another).

The weaker the first six regressions (no association or a negative association), and the stronger the last regression (positive association), the stronger is the claim that prostate cancer and female breast cancer have an unusually strong association, arguing for a common cause.

18.3.3   Looking at the Data

It is worthwhile providing initial plots of the data. In this case, an interesting feature appears for prostate cancer.

It is obvious from a casual inspection of Figure 18.4 that, since breast cancer is on the rise and ovarian and colorectal cancer are on the decline in the time period, there cannot be a positive association in two of the first three regressions (unless filtering drastically changes the apparent relationships seen here).

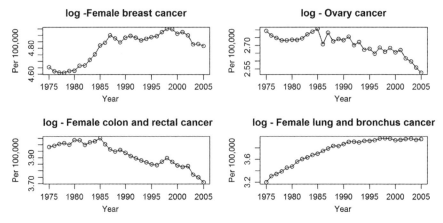

FIGURE 18.4   The trend, 1975–2005, for the four types of female cancer included in the study.

An unusual characteristic of the prostate cancer is that in 1986, the prostate-specific antigen (PSA) became available for early screening of prostate cancer. Because prostate cancer is a slow-acting cancer and might not otherwise be detected in early stages, this produces an artificial bump in prostate cancer incidence just after that time. In Figure 18.5, there is a sharp peak around 1992 in the prostate cancer graph due to this aberration. If there is a temporal association between breast and prostate cancers, this will almost certainly tend to dilute the effect. Whenever two variables are strongly associated, an external disruption to one of the variables will almost always dilute the apparent relationship between the variables.

Echoing general comments made when discussing the female cancers, the upward trend on prostate cancer and the downward trend on other cancers will almost guarantee that two of the three regressions involving male cancers have a negative association.

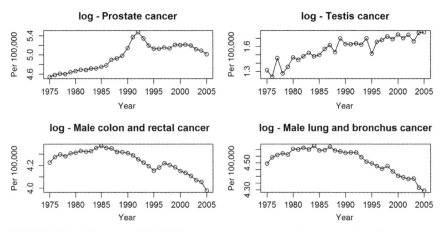

FIGURE 18.5   The trend, 1975–2005, for the four types of male cancer included in the study.

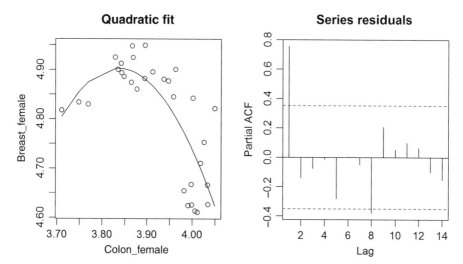

FIGURE 18.6 This model will require two filters, AR(1) followed by AR(2) for a valid regression. $R^2 = 0.5976$.

18.3.4   Examining the Residuals for AR(*m*) Structure

18.3.4.1   Displays of the Data   Each of the seven regressions is examined with regard to serial correlation in the residuals. A quadratic model was fitted for each regression, presuming this to be a saturated model. Plotting the data allows for a recognition of when a model is completely inappropriate. Overall, there seems to be little relationship between breast cancer and the cancers of ovary and colon, as the correlations are low and the relationships are primarily negative. However, filtering could always change matters.

In the past, the function *ar.yw()* has been used to analyze the residuals for the AR(*m*) order. In this section, because sample sizes are small, *ar.mle()* will be used. In an exercise, some of the problems with *ar.yw()*, for these small samples, will be highlighted.

Figures 18.6, 18.7, and 18.8 show the regressions on the unfiltered scale and the level of serial correlation in the data for the regressions between female breast cancer and the other three female cancers. Female breast cancer and lung cancer seem to be the only ones with a possible positive relationship. Recall that filtering will usually dilute, often substantially, the explained variation ($R^2 > R_F^2$).

Figures 18.9, 18.10, and 18.11 show much the same results for the relationships between the male cancers and the prostate cancer, as was seen with the regressions involving female cancers. The relationship between prostate and testis cancer, with locations in close proximity, is the only one that might have some positive association.

As anticipated, the regression between breast and prostate cancer, Figure 18.12, shows promise of a stronger positive association than the other regressions. Of course, the data must be filtered before any definitive conclusions are drawn.

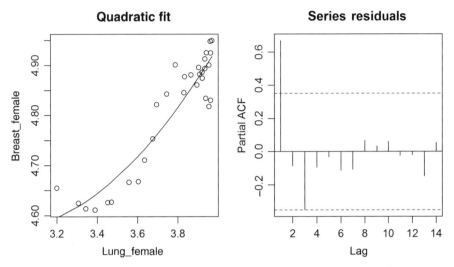

FIGURE 18.7   This model will require an AR(3) filter for a valid regression. $R^2 = 0.8649$.

18.3.4.2   Filtering Twice   The notion of filtering twice was first introduced and explained in Section 16.4.2.6. Unfortunately, small data sets are more prone to the need of more than one filter than large data sets. When multiple filters are applied, the effective sample size, n_F, which was already small, continues to diminish. Nevertheless, if the final vector $A_3A_2A_1y$ has quasi-independent observations, the results are valid on the filtered scale and produce information useful on the original scale.

An additional caution is required. If a data set is small, it is hard to pick up serial correlation. The apparent white noise in the final residuals is due to two elements: the

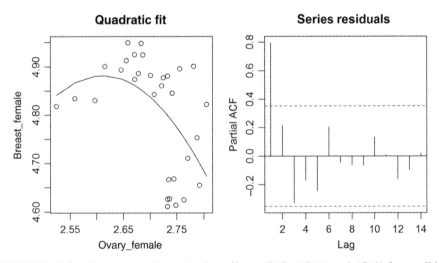

FIGURE 18.8   This model will require three filters AR(5), AR(1), and AR(1) for a valid regression. $R^2 = 0.2516$.

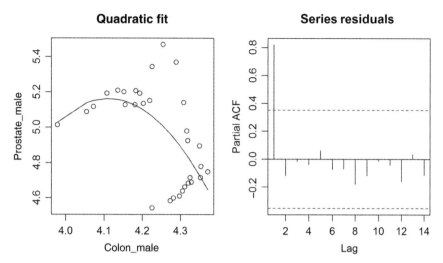

FIGURE 18.9   This model will require two filters AR(2) and AR(1) for a valid regression. $R^2 = 0.3465$.

filters are creating quasi-independence in the responses and the resulting residuals from a regression, and the sample size is shrinking, making deviations from white noise harder to detect.

Nevertheless, it does not make sense to just ignore small data sets, nor does it make sense to ignore the serial correlation. Small data sets have something to say, but even really large data sets can only say so much. It is the compilation of many data sets in many different contexts that provide strong support for theories.

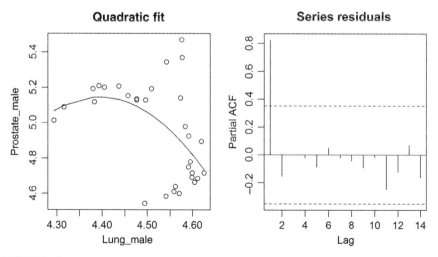

FIGURE 18.10   This model will require two filters AR(2) and AR(1) for a valid regression. $R^2 = 0.2489$.

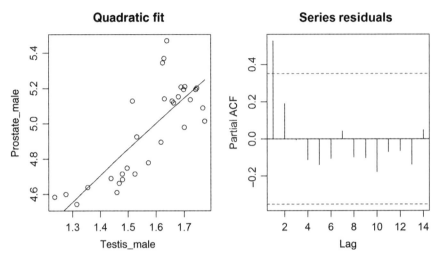

FIGURE 18.11   This model will require two filters AR(1) and AR(3) for a valid regression. $R^2 = 0.6434$.

18.3.5   Regression Analysis with Filtered Data

18.3.5.1   Results for the three Female Regression   A number of useful displays will now be introduced that will also be useful in later chapters and used for the rest of the book. The idea behind these displays is to communicate how the regression worked as a regression and as represented in a temporal order.

To most observers, the idea is that, when plots of x and y are placed side by side, in the order of time, they seem to line up visually. For a simple regression of x and

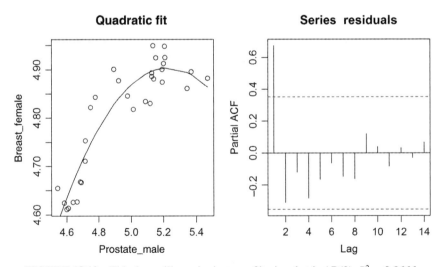

FIGURE 18.12   This data will require just one filtering that is AR(9). $R^2 = 0.8666$.

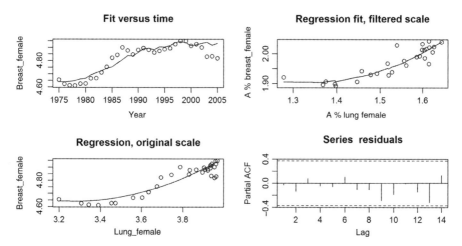

FIGURE 18.13 Female breast cancer versus female lung cancer, both variables on a log scale.

y, the regression algorithm finds the best values for a and b such that the line $a + bx$ closely matches the line y in time. If x and y move up and down together, the lines can be made to match closely using this method, but if the lines move in opposite directions, the result will be a line roughly of the form $a = \bar{y}, b = 0$.

However, time is not explicitly involved in the regression, except for the filtering of the residuals. A complete visual understanding of the modeling process should include: (i) the response y and the fitted line over time, (ii) the regression on the filtered scale, and (iii) the regression on the original scale. Because it takes no more space to plot four graphs than three, a fourth location can contain other important information: diagnostic plots or summary statistics. The fourth location is used here to display the *pacf()* plot for the residual from the filtered regression.

For Figure 18.13, the underlying model is log(*breast cancer*) ~ log(*lung cancer*) + log(*lung cancer*)^2. The code for the four plots is given below.

X is the usual regression matrix whose three columns are ones, log (*lungcancer*), and log(*lungcancer*)^2. A is a filtering matrix based on the residuals diagnosed as AR(3).

"final_fit" is the name of the final filtered regression fit.

Northwest panel of Figure 18.13:

```
# a plot of the response (breast cancer rates on a logarithmic scale) with the
# fitted filtered quadratic model
y_fit <- X%*%final_fit$coeff
plot(1975:2005,breast_female, xlab = "year") # log-breast cancer versus time
lines(1975:2005,y_fit) #predicted breast cancer over time, based on lung cancer
title("Fit versus time")
```

Northeast panel of Figure 18.13:

```
# the filtered x and y variables are plotted. A line for the fitted regression model is laid
on #this plot.
plot(A%*%lung_female, A%*%breast_female) # plot the filtered data
# for the line command below, a smooth curve requires that the x variable be placed
in numerical order, the variable "tmp" keeps the x and y fits properly linked.
tmp <- order(A%*%lung_female)
x <- A%*%lung_female
lines(x[tmp],final_fit$fitted[tmp])
title("Regression fit on filtered scale")
```

Southwest panel of Figure 18.13:

```
# similar to the previous plot, but on the unfiltered scale.
y_fit <- X%*%final_fit$coeff
plot(lung_female, breast_female)
tmp <- order(lung_female)
lines(lung_female[tmp],y_fit[tmp])
title("Regression fit on original scale")
```

Southeast panel of Figure 18.13:

```
pacf(final_fit$resid)
```

The northwest corner plot is the one most people want to see. It shows that the two variables can be made to fit very close to each other over time. However, the right-side plots show the performance of the regression fit—the quality of the regression on the filtered scale and the closeness of the residuals from this regression to white noise. It would not be too much of an over-simplification to say that folks interested in what the model is saying are interested in the graphs on the left-hand side; while those interested in a critical evaluation of the quality of the model are interested in the right-hand side graphs. It is left as an exercise for the student to explain why the fit on the original scale is a smooth quadratic, but the fit on the filtered scale has some rough corners.

If only two plots are displayed, the fit over time and the filtered regression provide the most information. As can be seen from Figure 18.14, breast cancer is increasing over time, while (female) colon and ovary cancers are dropping over time, so the best fitting regression line on the filtered scale is basically a flat line at the mean of filtered breast cancer (on the logarithmic scale).

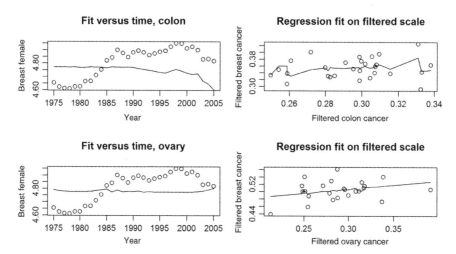

FIGURE 18.14   The relationship between female breast cancer and female colon cancer (upper plots) and female breast cancer and ovary cancer (lower plots).

18.3.5.2   *Results for the three Male Regressions*   In all three cases, the relationship is quite weak between prostate cancer and the other three male cancers (Figure 18.15) on the filtered scale (log transformation).

18.3.5.3   *Results for the Prostate Versus Breast Cancer Regression*   For the key result, it is best to show all four graphs (Figure 18.16). Although there is a problematic peak at around 1992 in the prostate cancer data, allowing a quadratic term in the model softens this peak a bit. Certainly, this regression has the strongest fit on the

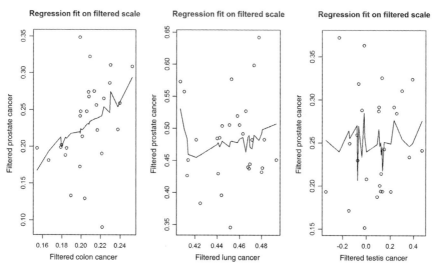

FIGURE 18.15   Filtered regressions for the other male cancers versus prostate cancer (log scale).

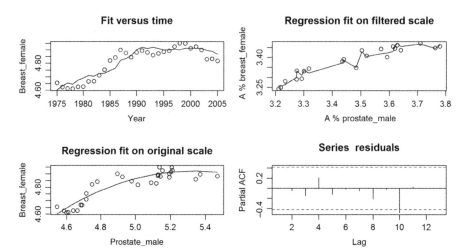

FIGURE 18.16   There is a close positive association between female breast cancer and prostate cancer.

filtered scale of any of the regressions. Although the focus has been on graphical analysis, Table 18.4 provides more numerical measures of strength of association. The value "naive R^2" in the table is just the explained variation for a quadratic model without filtering. This value provides a reminder that measures computed on serially correlated data can be very misleading.

It would be possible to perform some model selection and hypothesis tests, but the evidence is relatively clear from the graphs and the values of R_F^2 and $R_{F,pred}^2$. There is a very strong positive relationship between male prostate and female breast cancers. On the other hand, the only relationship in the other regressions that stands out is the breast cancer–lung cancer relationship in the females. This relationship is, however, noticeably weaker than the breast cancer–prostate cancer relationship.

Certainly, this data is essentially what the researchers wanted to see. Generally, cancers, even within a sex and in the same period and location, are relatively

TABLE 18.4   Measures of Strength of Association for the Seven Regressions

Regression	Naive R^2	R_F^2	$R_{F,pred}^2$	Filters	n_F
Female					
Breast vs. colon	0.5976	0.1182	0.0000	AR(1), AR(2)	28
Breast vs. lung	0.8649	0.8122	0.7478	AR(3)	28
Breast vs. ovary	0.2516	0.0929	0.0000	AR(5), AR(1), AR(1)	24
Male					
Prostate vs. colon	0.3465	0.1686	0.0278	AR(2), AR(1)	28
Prostate vs. lung	0.2489	0.2199	0.0784	AR(2), AR(1)	28
Prostate vs. testis	0.6434	0.0272	0.0000	AR(1), AR(3)	27
Breast vs. prostate	0.8666	0.9664	0.9581	AR(9)	22

uncorrelated, but breast cancer in females and prostate cancer in males go completely against this pattern. Recall, further, that it is the conjecture of the researchers that if breast cancer were separated into ER positive and ER negative, the relationship between ER positive breast cancer and prostate cancer would be even stronger.

The association between male prostate cancer and female breast cancer, given the weakness of the other relationships, does give one pause for thought, whatever the cause.

Of course, this is a long way from finding a common cause. The researchers have a theory that pollutants disrupt the endocrine system, causing specific types of cancers. This data is consistent with their conjecture, but there are many possible explanations for this link, and many potential data sets with which to compare competing explanations of the observed relationships. Remember, it took thousands of studies to establish that smoking causes lung cancer (see "Against All Odds" in the references).

Finally, the sample sizes are quite small, especially after filtering.

EXERCISES

1. Perform a nonparametric analysis of the first data set. Compare the results to the t-tests found in the text.

2. Using $ar.yw()$ for small samples: The purpose of this exercise is to show that the residuals are not filtered properly, when using $ar.yw()$ for a specific data set.
   (i) Fit the model log(prostate cancer) $\sim$ colon cancer + colon cancer^2 for males.
   (ii) Use $ar.yw()$ on the residuals to find values to build a filter.
   (iii) Create the filtering matrix A.
   (iv) Show $A(residuals)$ is not white noise.
   (v) Use $ar.mle()$ on the residuals and create a filtering matrix B.
   (vi) Show $B(residuals)$ is close to white noise.

3. For the female regressions:

   log(breast cancer) $\sim$ log(colon cancer) + log(colon cancer)^2
   log(breast cancer) $\sim$ log(lung cancer) + log(lung cancer)^2
   log(breast cancer) $\sim$ log(ovary cancer) + log(ovary cancer)^2

   Complete the following table to compare $ar.yw()$ and $ar.mle()$.

Regression	$ar.mle()$ order	$ar.mle()$ estimates $\hat{a}_1 \ldots \hat{a}_m, \hat{\sigma}_w^2$	$ar.yw()$ order	$ar.yw()$ estimates $\hat{a}_1 \ldots \hat{a}_m, \hat{\sigma}_w^2$
Colon	??	??	??	??
Lung				
Ovary	??			??

4. Verify the regression $\log(\text{breast cancer}) \sim \log(\text{colon cancer}) + \log(\text{colon cancer})^2$ must be filtered twice.

   (i) Fit a regression model to the data.

  (ii) Use *ar.mle()* to develop a filtering matrix A based on the residuals from this model.

 (iii) Fit a filtered regression to the data.

 (iv) Use *ar.mle()* to show these residuals appear far from white noise.

5. For the second data set, why are the quadratic models smooth fits on the original scale, but rough/choppy fits on the filtered scale?

6. Perform a complete analysis of the final regression (save all R code and turn it in).

   (i) Fit a model $\log(\text{breast cancer}) \sim \log(\text{prostate cancer}) + \log(\text{prostate cancer})^2$.

  (ii) Use the residuals to create a filtering matrix.

 (iii) Verify that *A(residuals)* is close to white noise.

 (iv) Fit a filtered model to the data.

  (v) Verify that the residuals from the filtered model are close to white noise.

 (vi) Construct the four plots from the text.

 (vii) Verify R_F^2 and $R_{F,\text{pred}}^2$

19

CHRISTOPHER TENNANT/BEN CROSBY WATERSHED DATA

19.1 BACKGROUND AND QUESTION

This data, due to graduate student Christopher Tennant and Professor Ben Crosby of the Idaho State University Geosciences department, is a measure of water flow rates at 12 watersheds in Idaho. The watersheds are contained within three different elevation ranges, representing different types of precipitation (low elevations have precipitation dominated by rain, higher elevations have snow-dominated precipitation, and the mid-elevations have mixed precipitation). The data is in the Tennant folder on the data disk (this data is not in DataMarket):

Low (rain) elevation: Baker Gulch, Gregory Creek, Rice Creek, Rock Creek.

Mid (mixed) elevation: Boulder Creek, Little Goose Creek, North Fork Slate Creek, Slate Creek.

High elevation (snow): Beaver Creek, Frenchman Creek, Salmon River, Smiley Creek.

The final measurement, which was taken for 1054 days (2.89 years) was normalized discharge (the volume of discharge per unit of time divided by the surface area of the watershed). The final units are millimeters per day. Normalized discharge reveals the amount of water contributed per unit area from a drainage basin and provides a convenient measure for comparing water flux from watersheds of different size.

Basic Data Analysis for Time Series with R, First Edition. DeWayne R. Derryberry.
© 2014 John Wiley & Sons, Inc. Published 2014 by John Wiley & Sons, Inc.

This is a preliminary analysis of the data and the interest is in whether the three regimes behave differently. In the main analysis, an exploration of whether the low and mid-elevations produce different patterns is undertaken. In the exercises, a similar analysis of the low versus high elevations is undertaken.

A couple of methods will be used to compare fit within groups to fit between groups. The general idea is that each watershed is different, but that if the classification into low, mid, and high elevations is important, the watersheds within a group will be more alike than they would be to watersheds in other groups.

19.2   LOOKING AT THE DATA AND FITTING FOURIER SERIES

19.2.1   The Structure of the Data

As with much of the data found in the real world, there is a distinct right skew, and a logarithmic transformation is desirable. In this case, the transformation is $z = \log_e(y + 0.001)$, where y values are the original measurements. In a couple cases, the smallest observation is 0, and the logarithmic transformation cannot be applied to 0. In order to apply the logarithmic transformation, an amount must be added to each observation—0.001 was chosen.

19.2.2   Fourier Series Fits to the Data

The first fit to the data will be a saturated model (365 days), so that the residuals can be diagnosed for AR(m) structure. For all parts of the analysis, the more alike watersheds within a regime are, the more promising the analysis. In this part of the analysis, it will be great if all watersheds, within an elevation class, have similar AR(m) structure.

The mid- and low-elevation fits look very different. The peaks in the mid-elevation watershed rates (Figure 19.1) are more symmetric, while the low-elevation watersheds have slow buildups to a peak, then sharp declines (Figure 19.2).

Three of the four low-elevation (Figure 19.2) watersheds had a very dry period in the first full low phase in the data.

Although the residuals from many of the sites are quite complex, as high as AR(30), the residuals from the regressions between sites might not display the same degree of complexity.

19.2.3   Connecting Patterns in Data to Physical Processes

Why are the residuals so complex? Although the complexity of the residuals is partly a subject matter issue, it behooves a consulting statistician to come to any meeting with subject-matter experts armed with some speculations, or at least some educated questions, as to why the residuals are so complex and as to why the residuals seem to be more complex in some cases than others. (A key characteristic of a good consultant is that they be willing to look stupid in the short run, in order to learn in the long run.)

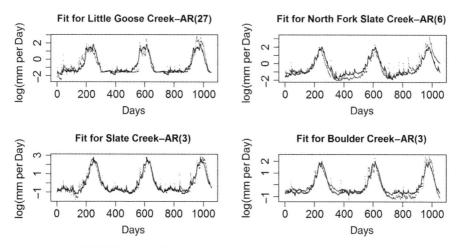

FIGURE 19.1   The initial fits for the mid-elevation watersheds.

Recall that residuals that are AR(m) suggest that, on average, the previous m residuals form patterns deviating from the signal in a systematic manner. Furthermore, river systems seem to have three components to the system: signal, noise, and random precipitation events that could be understood as random shocks that produce short-term deviations from both the signal and the noise. The statistician should be prepared to get these ideas out for consideration.

In this context, it may be reasonable for the statistician to add questions like the following: How often do precipitation events occur? How long would their impact be noticed in the environment? How might these impacts differ between the low- and mid-elevation systems? If a statistician works with a group over a period of time, that statistician should be able to develop a notion of what physical processes are driving

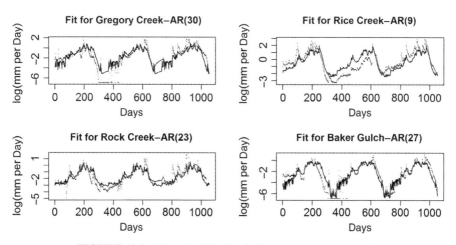

FIGURE 19.2   The initial fits for the low-elevation watersheds.

patterns in the data and should be able to contribute to the discussion in the language of the subject-matter experts.

19.3   AVERAGING DATA

To the extent that the goal is to compare the different systems, a working hypothesis, worth examining, is that the low systems are more like each other and less like the mid-elevation systems, and vice versa. In comparing the systems in Figure 19.3, something already noted is the qualitative differences between the peaks. Other differences in overall shape may be important as well as are the differences in timing of peaks and troughs.

For each watershed, a degree of fit *within* a watershed elevation group can be found by forming a regression of any one watershed against the average of the other three watershed values.

Analogously, for each watershed, a degree of fit *between* that watershed and the other elevation group can be found by forming a regression of that watershed against the average of all watersheds from the other group. This gives a rough numerical summary (R^2) and graphical display of the within-group and between-group concordance or discordance. A complete numerical summary appears in Table 19.1 and a graphical display using Gregory Creek is also given (Figure 19.4).

It is not hard to see that the fitting within a watershed group is much better than the fit between groups. As can be seen from Figures 19.1, 19.2, and 19.3, the fact that the watershed elevation groups have different overall shapes makes it impossible for the average of the mid-elevation watersheds to fit Gregory Creek well, although the average of the other three watersheds within the elevation group produce a quite reasonable model for Gregory Creek.

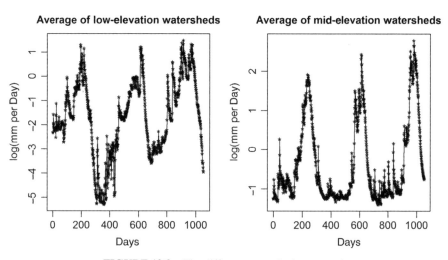

FIGURE 19.3   The different watersheds averaged.

TABLE 19.1   Preliminary Fit Measures Between and Within Groups (Fit to Original Scale, Based on the Model from Fitting a Simple Regression on the Filtered Scale)

Watershed	Fit within elevation groups R^2_{pred}, (R^2)	Fit between elevation groups R^2_{pred}, (R^2)
Low elevation		
Gregory Creek	0.8074	0.2200
	(0.8085)	(0.2224)
Rice Creek	0.5267	0.3334
	(0.5289)	(0.3354)
Rock Creek	0.6831	0.2842
	(0.6844)	(0.2866)
Baker Gulch	0.6585	0.3139
	(0.6597)	(0.3159)
Mid elevation		
Little Goose Creek	0.7553	0.2291
	(0.7564)	(0.2324)
N Fork Slate Creek	0.8302	0.4118
	(0.8307)	(0.4140)
Slate Creek	0.8067	0.1702
	(0.8075)	(0.1730)
Boulder Creek	0.7632	0.1902
	(0.7645)	(0.1933)

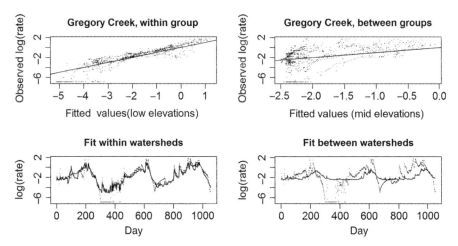

FIGURE 19.4   Using Gregory Creek as an example. The fit within a watershed and the fit between watersheds are compared (analysis on the filtered scale above, and original scale below).

19.4   RESULTS

It is quite obvious from a casual inspection of the data and preliminary graphs and computations that any analysis will show that the differences between groups are substantial compared to the differences within groups (exercises). Nevertheless, a definitive report of the findings should involve filtering the data and reporting the levels of R^2_{pred} on the original scale.

It is plain from a display of the relationships (Figure 19.4) that the fit within groups is much better than a fit between groups. In Figure 19.4, Gregory Creek is employed as an example. The fitted parameters are found using filtering, but the plots are of the data on the original scale. The upper graphs are the regressions and the lower graphs are the fitted values versus the observed values, over time. Numerical summaries save space and produce hard numbers. From Table 19.1, it is clear that the story is similar for all the watersheds.

Based on the results of Table 19.1, it is clear that every watershed fits (the average of) others within its elevation group far better than it fits the average of the other elevation group. Overall, the mid-elevation group seems to display more internal consistency than the low-elevation group, judging by the higher R^2_{pred} values within the group.

Paradoxically, North Fork Slate Creek has the distinction of having the highest correlation both to others within its group (0.8302) and to the average of the other groups (0.4118).

It is now time to ask the geoscience experts where they want to go next with this data.

EXERCISES

1. Perform a similar analysis comparing the high- and low-elevation groups.

   (i) Fit a Fourier series to each watershed and find the estimated AR(m) order of the residuals.

   (ii) Plot the averaged watershed value for the high-elevation groups and discuss how this plot compares to the other watersheds. Is there any sense in which the shapes of the three groups form a natural progression from low to high elevations?

   (iii) Form a plot similar to Figure 19.4 using Frenchman Creek.

   (iv) Form a table as in the chapter, but comparing the low to the high elevations.
      (a) Perform the analysis without filtering the data.

      (b) Perform the analysis filtering the data [parts (iii) and (iv) will be different].

   Observe that the patterns are the same in both tables (filtered and unfiltered), although the R^2 values are consistently higher for the unfiltered analysis. Explain why this pattern occurs.

20

VOSTOK ICE CORE DATA

20.1 SOURCE OF THE DATA

The Vostok ice core data is stored at the National Oceanic and Atmospheric Administration at these locations:

- ftp://ftp.ncdc.noaa.gov/pub/data/paleo/icecore/antarctica/vostok/deutnat.txt
- ftp://ftp.ncdc.noaa.gov/pub/data/paleo/icecore/antarctica/vostok/co2nat.txt.
  ftp://ftp.ncdc.noaa.gov/pub/data/paleo/icecore/antarctica/vostok/dustnat.txt

The original source (updated 11/2001) for this data is due to Petit et al. (1999).

The first file listed above contains a second column that is the estimated age of the reading and the last (fourth) column that is the temperature as deviations from a baseline level. The units for temperature are Celsius. A header in the original file explains each column in a lot more detail. Some of the first few rows of data in the file are presented in Table 20.1.

This data is in the Vostok folder from the data disk and is denoted "Vostok temps.txt." Obviously, the file must be brought into R using *read.table()* (with header = T), not *scan()*.

The second files from NOAA contains the ice core CO_2 measurements in ppmv (parts per million by volume). There are only two columns, the first is the estimated age and the second is the estimated CO_2 level. The first few rows are presented in Table 20.2.

Basic Data Analysis for Time Series with R, First Edition. DeWayne R. Derryberry.
© 2014 John Wiley & Sons, Inc. Published 2014 by John Wiley & Sons, Inc.

TABLE 20.1   Temperature Data ("Vostok temps.txt" in the Vostok Folder)

Depth_corrected	Ice_age(GT4)	deut	deltaTS
0	0	−438	0.00
1	17	−438	0.00
2	35	−438	0.00
...			
10	190	−435.8	0.36
11	211	−443.7	−0.95
12	234	−449.1	−1.84
13	258	−444.6	−1.09
14	281	−442.5	−0.75, etc.

These are obviously not direct measurements, and some investigation at the site would be required to determine the exact process whereby these values were estimated. Since this author does not have sufficient scientific background to assess the methodology by which changes in temperature, age, and/or CO_2 levels were estimated, the numbers will be taken at face value here.

Ignoring the context of the problem, it would be hard for anyone to look at these graphs aligned in time (Figure 20.1) and not think there is something here requiring explanation.

20.2   BACKGROUND

It has been noted that both CO_2 levels and temperature have been rising in recent years. There is a strong association and a potential causal mechanism (the greenhouse effect). However, the relationship could be strong (CO_2 drives temperature) or weak (the greenhouse effect could be minor and not the main reason for temperature change).

Perhaps the strongest counterargument to the climate change argument is that climate changes all of the time due to many, as yet poorly understood, causes. For example, there were a number of very cool periods (little ice ages), alternating with warming periods between 1550 and 1850 (see "Little Ice Age" in Wikipedia for extensive references).

TABLE 20.2 CO_2 Data ("Vostok CO2.txt" in the Vostok Folder)

Gas_age	CO_2
2342	284.7
3634	272.8
3833	268.1
6220	262.2
7327	254.6
8113	259.6, etc.

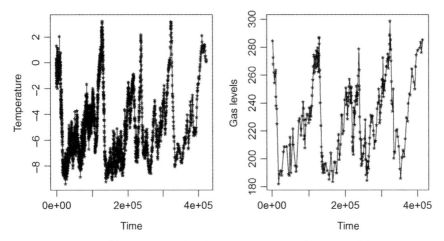

FIGURE 20.1   The parallels in these graphs are, of course, striking.

The best evidence for countering this kind of criticism of a link between CO_2 and temperature would be based on evidence showing a strong association over a very long time period (so long that long cycles of climate change could not be a highly plausible alternative explanation).

The Vostok ice core data goes back a little over 400,000 years. The data provides 283 measurements of CO_2 and 3331 measurements of temperature change over this period. How strong is the association between CO_2 levels and temperature change in this data? Using this data, is it possible to estimate the typical temperature change associated with a fixed increase in CO_2 levels?

Disclaimer—please read before proceeding: Throughout this analysis the use of causal language will be avoided, not because of any skepticism about climate change, but because a single data set, based on an observational study, should never be used as the basis for a causal argument. The inference of cause–effect is a scientific pronouncement based on (i) all available evidence, (ii) the potential causal mechanisms, and (iii) the relative plausibility or implausibility of alternative explanations. In this regard, the video "The Question of Causation"(Against All Odds in the references), which follows the long line of evidence leading to the claim "smoking cause lung cancer", would be worth viewing. This will be analyzed as a single data set; it is up to others to put all the evidence together. As a general practice, statisticians always avoid the use of causal language when discussing observational studies. Any assessment of causality in such cases is in the realm of subject-matter experts and not statisticians (or spin doctors).

20.3   ALIGNMENT

20.3.1   Need for Alignment, and Possible Issues Resulting from Alignment

The ice cores yield 3331 observations of changes in temperature, while the CO_2 data yields 283 observations over the same roughly 400,000 years. In order to get a

TABLE 20.3   The Aligned Data

Ice_age	deltaT	Gas_age	CO_2
2331	−0.98	2342	284.7
3646	0.65	3634	272.8
3824	−0.46	3833	268.1
6241	0.09	6220	262.2
7315	−0.29	7327	254.6
8135	2.06	8113	259.6
10,124	−0.58	10,123	261.6, etc.

workable data set, the 283 CO_2 observations must be matched with 283 of the 3331 temperature observations that match closely in time.

Alignment of the data was done manually by the author. It was based on finding close matches (in time) of the 3331 temperature values to the 283 CO_2 values. Because two data sets have been aligned to produce a new data set, a number of questions needed to be addressed in order to continue with the analysis:

(i) Does the substantially reduced set of temperature data (283 vs. 3331 observations) still have the same basic pattern over, time, as the original data?

(ii) Are the years for the CO_2 and temperature data closely matched?

(iii) To what degree are observations equally spaced? All the analyses we have done have been on equally spaced time series; this will no longer be the case (at least not exactly).

The aligned data is in the Vostok folder of the data disk and denoted as "aligned.txt." The first few rows are given in Table 20.3. The degrees of match between columns 1 (the estimated time of the temperature measurement) and 3 (the estimated time of the CO_2 measurement) is the first concern.

20.3.2   Is the Pattern in the Temperature Data Maintained?

To what extent this is to be expected and to what extent its sheer luck is unclear, but fortunately the subset of 283 observations seems to have kept the fundamental patterns in the data intact (Figure 20.2). In fact, many detailed features of the overall pattern are preserved in the subsample.

20.3.3   Are the Dates Closely Matched?

The times are well matched. The largest single difference was 317 years. About half the differences are less than 30 years. Given that the times span from about 2331 to 41,410, these are small differences. Based on the left panel of Figure 20.3, it would be hard to expect a better match between times than was found.

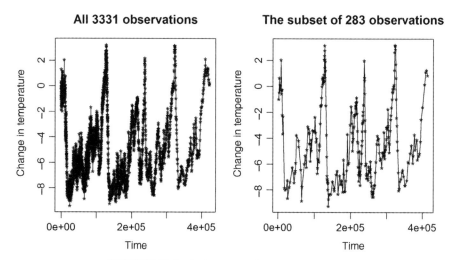

FIGURE 20.2   Is the temporal pattern preserved?

20.3.4   Are the Times Equally Spaced?

A variable, denoted "lag," can be computed by taking differences between adjacent time periods. If this is done for the time stamps for the temperature data, the results are as follows summary(lag):

Min.	1st Qu.	Median	Mean	3rd Qu.	Max.
55.0	665.8	1018.0	1460.0	1926.0	6012.0

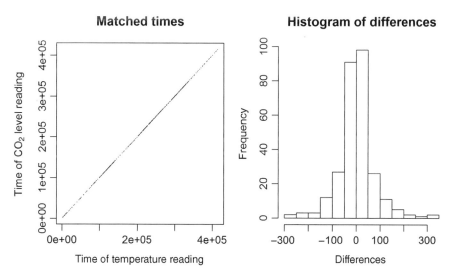

FIGURE 20.3   A scatterplot of time of CO_2 reading versus time of related temperature reading, and a plot of the differences, in time, between related readings.

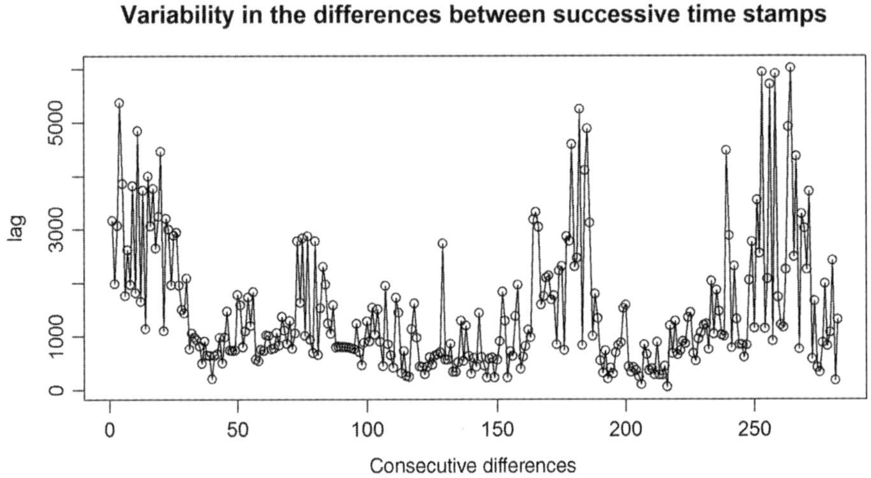

Variability in the differences between successive time stamps

FIGURE 20.4   The differences between time stamps over time.

Obviously, this will be the most significant challenge to analyzing this data as a time series. Two different analyses will be pursued. In the first approach, the data will be naively analyzed as if the spacing were equal. In the second analysis, an approach assuming an AR(1) model will be used. The AR(1) model can be generalized directly to unequal spacing. It should not be expected that either approach will be fully satisfactory, but both approaches should yield some insights into the data.

Figure 20.4 shows that there is no particular temporal pattern to the unequal spacing, in other words, large and small time lags are scatter throughout the entire time period.

20.4   A NAÏVE ANALYSIS

20.4.1   A Saturated Model

The naïve analysis is quite straightforward. What could go wrong? In this model CO_2 levels will be used to predict temperature. A cubic fit to the data will be chosen initially, although it would be nice to be able to reduce to a linear or quadratic model. The criteria $R^2_{F,pred}$ will be used for model selection.

Because the CO_2 values are quite large, a new variable, based on "CO_2–mean (CO_2)," (the mean is 234.0), will be used. This will prevent the squared and cubic terms in the regression from becoming excessively large and will reduce problems of co-linearity in the matrix of X values.

Fitting a cubic model, Figure 20.5, produces residuals that are AR(2).

Furthermore, although AR(2) was selected, models of order 4 and 5 are very close in AIC value (Table 20.4). It may be wise to immediately adopt an AR(5) model to assure complete filtering.

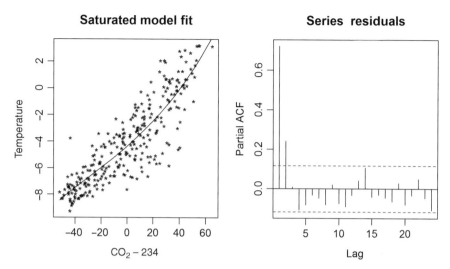

FIGURE 20.5   The initial saturated fit and residuals from the naïve (ignoring unequal spacing) analysis.

As first discovered in Section 16.4.2.6, sometimes the data must be filtered twice. If an AR(2) model for the error were adopted here, this would be required. A way to avoid this, at least sometimes, is to foresee that a more complex model for the residuals looms on the horizon. The purpose of the filter is to get quasi-independent observations without giving up much sample size. The filter itself is of no intrinsic interest, nor is the order itself important. In this particular case, where the spacing is not even equal, it is particularly hard to attach any importance to the AR(m) model itself. In cases like this using a higher order filter, when the decision is close, is like washing with extra strength detergent.

Two key factors here are: (i) that the order 4 and 5 models are within one unit of the best model in terms of AIC differences, a very small difference, and (ii) the estimated values for $\hat{a}_1 \dots \hat{a}_m$ are quite stable across all of these models (suggesting these more complex models really are refinements of, and not dramatically different from, the order 2 model).

TABLE 20.4   Several Models Fitted to the Residuals

Order	ΔAIC	$\hat{a}_1 \dots \hat{a}_m$
1	14.95	0.721
2	0.00	0.547, 0.241
3	1.98	0.545, 0.237, 0.008
4	0.70	0.546, 0.262, 0.067, −0.107
5	0.84	0.537, 0.267, 0.088, −0.063, −0.081
6	2.54	0.534, 0.265, 0.091, −0.054, −0.064, −0.033

TABLE 20.5   Model Selection Table ($n_F = 277$, $SST_F = 342.25$)

Model	p	SSE_F	R_F^2	$R_{F,pred}^2$	AIC_F
One mean	1	342.25	NA	NA	1620.45
Linear	2	225.10	0.3423	0.3300[a]	1506.38[a]
Quadratic	3	225.05	0.3425	0.3213	1508.32
Cubic	4	222.85	0.3489	0.3192	1507.60

[a]Best model.

20.4.2   Model Selection

After filtering once with an AR(5) filter, the resulting residuals are still AR(1) with $\hat{a} = 0.1355$. The second filter produces residuals not distinguishable from white noise [beginning with an AR(2) filter would have surely been even more convoluted].

Model selection information contained in Table 20.5 suggests that the linear model fits the data well, although not much better than the quadratic or cubic model. Because there is an interest in interpreting rates of change, it will be nice if a linear model holds up, suggesting a constant rate of change.

20.4.3   The Association Between CO_2 and Temperature Change

From a summary of the best fitting filtered model, Table 20.6, it is trivial to determine that an increase of one unit in CO_2 levels (ppmv) is associated with an increase of 0.059 in temperature (°C). The confidence interval is about a 0.0488 to 0.0684 increase in temperature (°C). It should be noted, before using the results from Table 20.6, that the residuals from the filtered analysis were assessed as white noise using *ar.yw()*.

Although the relationship is not really strong, based on $R_{F,pred}^2$, the mesh between temperatures and CO_2 levels is clear in the plot of "predictions versus time." Recall the difference between the model fit over time and the complete prediction over time, is that the predictions use the fit and the previous 6 residuals [because the model for the noise is AR(6)], to produce an overall forecast for the next time period. The combined adjustment for the two filters can be extracted directly from the first row of the matrix *A2%*%A*, the resulting filtering matrix after application of each filter.

TABLE 20.6   Summary of the Best Fitting Model (with Filters *A* and *A2*)

	Coefficients:			
	Estimate	SE	*t*-value	Pr($>|t|$)
A2 %*% (A %*% X) ones	−4.084623	0.249854	−16.35	<2e−16
A2 %*% (A %*% X)new_CO2	0.058579	0.004896	11.96	<2e−16

TABLE 20.7   Fit Between the Data and the Model on the Original Scale

Predictive model	R^2	R^2_{pred}
Fit alone	0.6491	0.6432
Fit with AR(6) adjustments	0.9129	0.9117

The quality of fit for both the fitted model and the predictions is given with a series of related R^2-like values on the original scale.

The fit of the data to the model, after adjusting for AR(6) noise (Table 20.7), is quite strong (Figure 20.6). The relationship may or may not be causal, but it is certainly not likely to be coincidental.

For the "fit alone" line, R^2_{pred} was found by combining the "hat" value from a linear fit on the original scale with the residuals from the model fitted on the filtered scale. These values were also used for the model with AR(m) adjustments, although it is not quite clear what "hat" values should be used in this case.

What follows in the rest of the chapter are attempts to "dig deeper" to assess both the reliability of this analysis and whether another analysis might yield better results. As a spoiler alert, there will not be much reason to either doubt or alter this analysis.

20.5   A RELATED SIMULATION

20.5.1   The Model and the Question of Interest

How might unequal spacing impact an analysis? Certainly this is a nebulous question, but a simulation meant to mimic this kind of data might offer some insights.

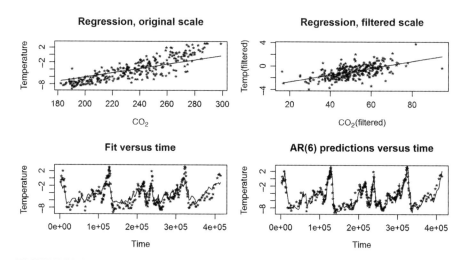

FIGURE 20.6   Displays for the naïve (ignoring unequal spacing) analysis of the Vostok data.

The following simulation will create data in much the way the Vostok data might be thought of as coming about and follow the same pattern analysis to get final results when it is known what the final result should be.

The scenario:

(i) An x variable varies over time.
(ii) A model of the form $y = \beta_0 + \beta_1 x + \varepsilon$.
(iii) An MA(3) model for the errors (ε) with $R^2 \approx 0.75$.
(iv) Create $n = 1000$ observations, but drop 71.7% of the observations at random.

The last step will create large gaps in the data producing data that is very unequally spaced in time. Three models will be compared: the true model, a model for all 1000 observations, and a model for the 283 observations left after dropping observations at random. The precise value of R^2 is unlikely to matter, but the goal is to have a model that has both a clear signal and a clear noise. A simulation with $n = 3200$ is left as an exercise. Although a larger simulation is closer to modeling the actual Vostok data, the graphs are very noisy, so the smaller simulation is used for illustration and the larger simulation is left as an exercise.

The reason the question being addressed is nebulous, whether analysis of unequally spaced time series can get reasonable results using this naïve approach, is no one simulation could be conclusive and it is hard to list a set of scenarios that would satisfactorily answer the question. In an exercise, the student will be asked to produce a couple of other scenarios (with $n = 3200$). As with other simulations in this book, this example is illustrative rather than exhaustive.

20.5.2   Simulation Code in R

The data in Figure 20.7 was produced with the following code:

```
time <- c(1:1000)
# create a different model for different periods of time
x <- rep(NA,1000)
x[1:250] <- 3*cos(2*pi*(time[1:250]/80 +0.1))
x[251:500] <- 5*cos(2*pi*(time[251:500]/115 +.15))
x[501:750] <- 2*cos(2*pi*(time[501:750]/123))
x[751:1000] <- 3*cos(2*pi*(time[751:1000]/75 + .35))
# this is a model from 15.1.3, the standard deviation is altered by trial and error until the\
# R2 value is about 0.75
error <- arima.sim(n = 1000, list(ma=c(0.8. 0.12. -0.144)), sd = 11.0)
y <- 3+10*x + error
fit <- lm(y ~ x)
plot(time,y,pch="*")
lines(time,fit$fitted)
title("The initial data")
```

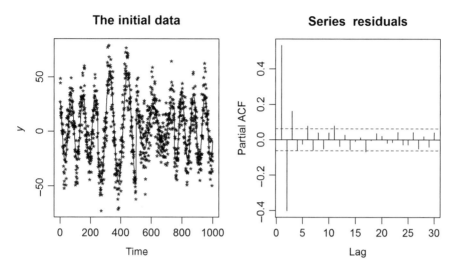

FIGURE 20.7   The original data for the simulation.

20.5.3   A Model Using all of the Simulated Data

The important things that were computed in the Vostok data were the confidence interval for the slope and the measures of fit and predictive accuracy on the original scale. The four-in-one graphs were also of value. All of these will be produced both for the full model and the model with just 283 observations. It is hoped the full analysis and the analysis with just 283 observations will be similar in these important ways.

The initial error structure appears quite complex, based on *ar.yw():*

1	2	3	4	5	6	7	8
0.7896	-0.4765	0.1698	-0.0195	-0.0744	0.1478	-0.1085	0.0460

Order selected 8 sigma^2 estimated as 112

The filter produced residuals estimated to be white noise using *ar.yw()* on the first filtering. The parameter estimates are quite good:

Coefficients:

	Estimate	Std. Error	t value	Pr(>ltl)
A %*% Xones	2.8659	0.6382	4.491	7.93e-06
A %*% Xx	10.3149	0.2558	40.329	< 2e-16

Using hypotheses tests as a reality check:
$(2.8659 - 3.0)/0.6382 = -0.162$ for the intercept and $(10.3149 - 10)/0.2558 = 1.231$. So the estimates match the parameters quite well (although the only direct interest is in the estimated slope, there would be some indirect concern if the estimate of the intercept were poor).

TABLE 20.8   **The Fit for the Filtered Model, the Original Scale, and the Original Scale with AR(8) Corrections**

Model	R^2	R^2_{pred}
Filtered	0.6216	0.6201
Fit, original scale	0.7739	0.7730
Fit with AR(8) adjustments	0.8640	0.8612

As usual, tracking and adjusting for AR(m) features in the residuals produce a better fit to the data (Table 20.8). As before, the "hat" values for the two models are found by fitting the model on original scale. The fit of the model to the data is displayed graphically in Figure 20.8.

20.5.4   A Model Using a Sample of 283 from the Simulated Data

Before examining a subset of the data, it is wise to anticipate what to expect. Comparing expectations to results is always more productive than just looking at results. It is hoped the slope is still estimated to be about 10.0, but the confidence interval should be wider.

It is hard to imagine what the filtering will be like, but it is expected that the R^2-like values will drop across the board.

The following steps produce a random sample of 283 of the time periods:

```
new_time <- sample(time,283)
new_time <- sort(new_time)
```

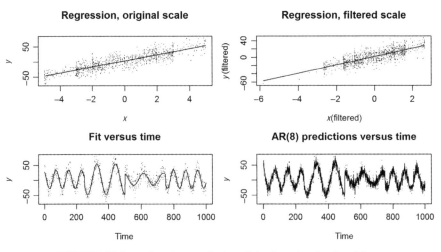

FIGURE 20.8   A four-in-one display of the fit to the simulated data.

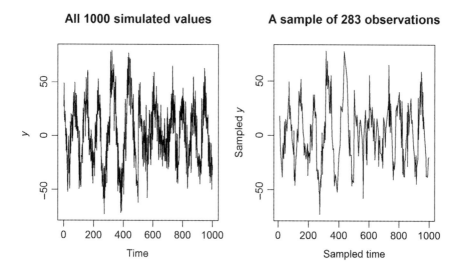

FIGURE 20.9   The subset of the original data still has much the same shape.

The sample observations appear to retain the general pattern of the entire data (Figure 20.9). The sampling scheme did produce erratic spacing, but not as unusual as found in the Vostok data.

A variable "lag" was defined as lag <- new_time[2:283] - new_time[1:282] with results, summary(lag):

Min.	1st Qu.	Median	Mean	3rd Qu.	Max.
1.000	1.000	2.000	3.532	5.000	14.000

A total of 75 of the lags were equal to one.

The data, after fitting a regression as shown in Figure 20.10, where new_y is y[new_time], now appears to be AR(1)! 0.1798 ,   Order selected 1 sigma^2 estimated as 172.8. Of course, any subsampling is expected to reduce the order, *m,* of the AR(*m*) model (why?).

The filtered model has residuals identified as white noise based on *ar.yw( )*.

	Estimate	Std. Error	t value	Pr(>|t|)
A %*% new`Xones	3.0378	0.9545	3.183	0.00162
A %*% new`Xnew`x	9.8493	0.3868	25.462	< 2e-16

The parameter estimates still seem reasonable. That is, $(3.0378 - 3.0)/0.9545 = 0.0396$ and $(9.8493 - 10.0)/0.3868 = -0.390$. Of course, the best way to assess this would be to fully automate the filtering process and perform this entire simulation perhaps 10,000 times and assess whether the results are as expected, in the long-run. The purpose of this section is more an exploration than a demonstration. As before, the filtered residuals are not much different from white noise based on *ar.yw( )*.

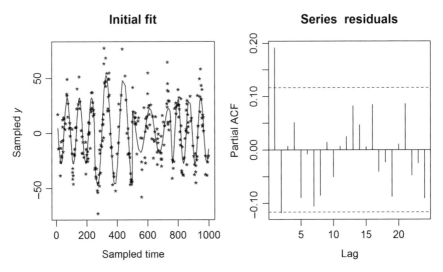

FIGURE 20.10   The initial fit to the data and the *pacf()* plot.

The model seems to fit the sampled data well (Figure 20.11). Because the corrected predictions only provide a small AR(1) correction, the fitted values and the predicted values are not much different.

The evidence from Table 20.9 is generally in agreement with Figure 20.11 that the filtering has become less effective. Comparing the fits to the full 1000 observations, there is less filtering [AR(8) vs. AR(1)], and while the filtered model has higher R^2 values with the subset of the data, this does not translate to better fit on the original scale.

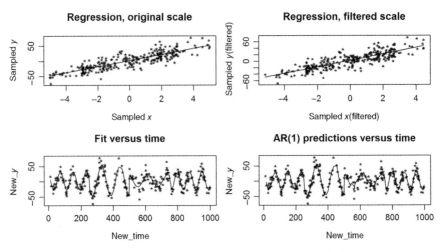

FIGURE 20.11   The fitted model using just $n = 283$.

TABLE 20.9   Measures of Fit for the Subset of 283 Simulated Observations

Model	R^2	R^2_{pred}
Filtered	0.6984	0.6941
Fit, original scale	0.7638	0.7605
Fit with AR(1) adjustments	0.7715	0.7666

Compared to the full sample, this model has slightly lower R^2 values on the original scale and these values do not improve nearly as much when adjustments are made to the residuals.

All of these results are good, considering the big picture. The implication is that, if a fuller set of data with equal spacing had been available, the confidence interval for the slope would have been narrower, and the R^2 values would have been higher. Although any single simulation carries limited weight, it does lend some credence to the confidence interval for the slope given earlier (in the analysis of the actual Vostok data). All of these support, in a very limited way, that the results of the naïve analysis may well be fully reasonable.

20.6   AN AR(1) MODEL FOR IRREGULAR SPACING

20.6.1   Motivation

It has been established that, for the AR(1) model, the model can be "re-scaled" in the sense that $\varepsilon_n = a\varepsilon_{n-1} + w_n$ implies, by iteration of the relationship, $\varepsilon_n = a^k \varepsilon_{n-k} + \sum_{j=0}^{k-1} a^j w_{n-j}$ (Chapter 5). The skip method discussed in Section 13.7.2, for example, relies on the fact that, if equally spaced observations have an AR(1) error structure with coefficient a, then observations taken by skipping k observations apart will have an AR(1) structure with coefficient a^k.

Modeling the error for one missing observation also reveals a pattern. Suppose the original errors are of the form: $\varepsilon_2 = a\varepsilon_1 + w_2$, $\varepsilon_3 = a\varepsilon_2 + w_3$, $\varepsilon_4 = a\varepsilon_3 + w_4$, $\varepsilon_5 = a\varepsilon_4 + w_5$, etc. However, if observation 3 is missing, the remaining errors could be modeled as: $\varepsilon_2 = a\varepsilon_1 + w_2$, ε_3 missing, $\varepsilon_4 = a^2\varepsilon_2 + aw_3 + w_4$, $\varepsilon_5 = a\varepsilon_4 + w_5$, etc. In other words, the model generalizes to an AR(1) model where the coefficient is $a^{\Delta t}$, with a further caveat that the variance is greater when Δt is large (in an exercise you will show the ratio of the largest to smallest variance for data with missing values, when the original data had equal variance, is at most $1/[1 - a]$).

Transitioning to a continuous model where the spacing is irregular, the model is $\varepsilon_t = a^{\Delta t}\varepsilon_{t-\Delta t} + \omega_{\Delta t}$, where $\omega_{\Delta t}$ is a combination of previous and current white noise components.

Using this idea, the parameter a for an AR(1) model can be estimated by compensating for the unequal time lags. A filtering loop can be derived based on Δt and $\hat{a}$. This may or may not provide an effective filter, depending largely on whether the AR(1) model with constant a really does fit the data to some degree. In our data, the

Δt's (lag_age) are quite large, we will find it is helpful, in solving the problem with the R function *nls()* to get manageable values. The normed value $\Delta m = \Delta t / mean(\Delta t)$ will be used. Summary statistics for Δm in the Vostok data.

Min.	1st Qu.	Median	Mean	3rd Qu.	Max.
0.03356	0.45720	0.68480	1.00000	1.26800	4.11200

20.6.2   Method

The goal, then, is to find an optimal solution to the (nonlinear) least squares problem, when the parameter a is allowed to vary: $\min \sum(\hat{\varepsilon}_j - a^{\Delta m_j}\hat{\varepsilon}_{j-1})^2$. Implemented in R this is

```
fit_initial <- lm(temperature ~ CO2 )
a_guess <- 0.5
next_residual <- fit_initial$residual[2:n]
previous_residual <- fit_initial$residual[1:(n-1)]
fit_a_nls <- nls(next_residual ~ a^(delta_m)*previous_residual, start=list(a = a_guess))
```

The estimated value, $\hat{a}$, using this code is 0.74694. By trying several starting values, it is easy to see that the final estimate, $\hat{a}$, is not dependent on the starting guess. The filtering can be handled using a "for" loop much more easily than a filtering matrix:

```
x_filtered <- rep(NA, 282)
y_filtered <- rep(NA, 282)
ones <- rep(NA,282)
CO2 <- CO2 - mean(CO2)  # in the previous analysis this was done to control
# colinearity. This is done here also, so the outputs
# are comparable.
for(j in 1:282)
{ x_filtered[j] <- CO2[j+1] - .74694^(delta_m[j])*CO2[j]
ones[j] <- 1- 0.74694^(delta_m[j])
y_filtered[j] <- temperature[j+1] - 0.74694^(delta_m[j])*temperature[j]}
x_filtered <- cbind(ones,x_filtered)
fit_filtered <- lm(y_filtered ~ -1+ x_filtered)
```

20.6.3   Results

The fitted model is summary(fit_filtered):

Coefficients:

	Estimate	Std. Error	t value	Pr(>|t|)
ones	-4.510897	0.198124	-22.77	<2e-16
x_filtered	0.057020	0.004627	12.32	<2e-16

Although the parameter estimates are reasonable (compare to Table 20.6), the resulting residuals, diagnosed as AR(3) using *ar.yw()*, are not close to white noise.

Also the standard errors are (slightly) smaller than the previous analysis of Table 20.6, in other words they are probably too small. On the other hand, the mean square error for this model is 0.8373348 (summing the squared residuals and dividing by 280, the degrees of freedom) versus 0.7975, the estimated variance of a fully filtered model [also from *ar.yw()* applied to these residuals].

A second filtering, assuming equal spacing, produces nearly identical results to the first approach, with residuals not distinguishable from white noise, based on *ar.yw()*:

Coefficients:

	Estimate	Std. Error	t value	Pr(>|t|)
A2 %*% Xones	-4.529512	0.250955	-18.05	<2e-16
A2 %*% Xx_filtered	0.050532	0.004916	10.28	<2e-16

MSE for this model being 0.77606.

Overall, this approach seems to support the previous analysis. In other words, this approach, followed up with a second AR(3) filter, got results similar to the first analysis. This analysis, while it did not fully succeed, would seem to be less dependent on the equal spacing assumption.

Usually, when two different, reasonable analyses yield nearly identical results for a set of data, the suggestion is that "All roads lead to Rome", and any reasonable analysis would produce the same result.

20.6.4   Sensitivity Analysis

Nevertheless, because this data is relatively famous (or infamous in some circles) and because both analyses made use of some key assumptions not met (the first analysis assumed equal spacing, which is certainly not close to being true; the second analysis assumed an AR(1) model and needed to use the assumptions of equal spacing for the second, admittedly, minor filtering), the data will be further prodded and poked without much additional insight.

It is possible that the model is AR(1), but that the value of *a* changes over the course of time. In any case, something could be learned from fitting the model over different subsets of the data. Dividing the data into four segments of roughly 70 data point, the filtered regressions were run after using *nls()* to get a different *â* for each period.

The analysis with varying time periods (Table 20.10) suggests the value of *a* does vary over time, but that even reducing the time periods to about 100,000 years allows, half the time, for an AR(1) model to filter the data such that the residuals cannot be distinguished from white noise using *ar.yw()*. Although there is variation in the parameter estimates, value similar to the previous two analyses is found.

TABLE 20.10   Analysis over Different Time Periods

Model	n	$\hat{a}$ (SE)	$\hat{\beta}_0$ (SE)	$\hat{\beta}_1$ (SE)	Order of residuals
10,151.1 years	71	0.6853	−4.13	0.060	0
(2336.5–12,487.6)		(0.088)	(0.396)	(0.0107)	
93,935.5 years	70	0.522	−4.45	0.0898	0
(12,575.5–219,709)		(0.125)	(0.25)	(0.0082)	
72,307.5 years	72	0.7305	−5.48	0.0479	2
(220,174.5–292,482)		(0.1007)	(0.43)	(0.0097)	
120,388.5 years	70	0.7975	−3.51	0.0481	2
(293,727.5–414,116)		(0.0784)	(0.42)	(0.0086)	

Parameter estimates are given with standard errors in parentheses on the next line.

20.6.5   A Final Analysis, Well Not Quite

The technique of moving windows could be used to estimate $a(t)$, the AR(1) parameter as a function of time. The data could then be filtered based on both the estimate $\hat{a}(t)$ and the spacings Δt.

It is possible to use moving windows to estimate a locally, but the *nls()* function experiences failure to converge for windows less than 47. Because of a desire to get highly variable local estimates, this window was used (the usual tradeoffs between bias and variance are at work in choosing a window here). Figure 20.12 shows the values of $\hat{a}$ for the narrowed possible window ($w = 47$) and a very wide window ($w = 95$).

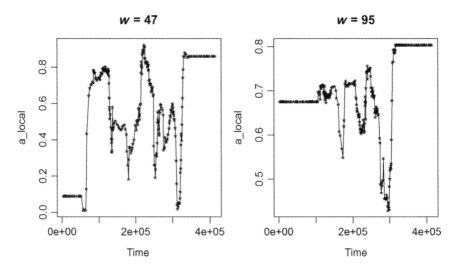

FIGURE 20.12   The local value of $\hat{a}$ is obviously smoother for a wider window.

Complexity of residuals over time

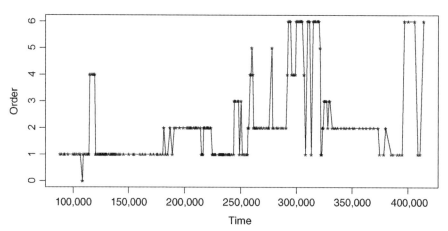

FIGURE 20.13   Local clusters of residuals, taken 40 at a time, with *ar.yw()* used to diagnose order, can have an order as high as 6.

These models produce regressions similar to those already discussed, but still with residuals that are not white noise. It is worth examining the residuals more closely.

Figure 20.13 shows that locally the order of the residuals can be anywhere from order 0 to order 6 as diagnosed using *ar.yw()*, although the usual caveat that the *ar.yw()* function assumes equal spacing.

It seems clear that none of the models considered in this last effort, assuming AR(1) errors, fully solve the problem of unequal spacing. However, a model for a more complex model, AR(2) or higher, with extremely unequal spacing is well beyond the scopes both of this book and the author's competence.

Nevertheless, the analyses provided here, which are substantial improvements on any method that ignores serial correlation, do produce relatively consistent results suggesting an increase of CO_2 of about 10 ppm would be associated with (this language is commonly used in the regression literature and is explicitly meant to avoid coming down on either side of the question of causation) an increase in temperature of about 0.5°C.

20.7 SUMMARY

A naïve analysis of the data, applying the modeling efforts from previous chapters while ignoring the unequal spacing of the observations produced an "acceptable" result. The result was acceptable in the sense that the filtering mechanism developed was successful in producing quasi-independent observations, the goal of filtering, even though the conditions assumed in developing the filter were not met.

A second simulation analysis suggested that this might not be an accident. In a simulation producing data with unequal spacing, it was found that the filtering can indeed work and the resulting statistical analysis seemed to match the model that produced that data.

Finally, another approach was explored that turned out to be a partial dead end, at least in this case. It was assumed that the data was AR(1) and a model was fit that allowed for varied spacing between observations. While this approach did not fully filter the data, a final AR(1) filter applied to the resulting filtered values produced a model similar to the naïve analysis.

However, further examination shows that an AR(1) is probably always going to look promising, but never quite solve the problem. This could be viewed as homage to all the dead ends that were explored in this book that ended up in the wastebasket, as well as all the dead ends consulting statisticians, researchers, and graduate students go down every day. Science only appears in the final publications and other final results. Behind every good result are a lot of full recycling bins. Certainly this was the most interesting dead end explored in the course of writing this book.

EXERCISES

1. Filter the Vostok cubic regression with an initial AR(2) model, the one suggested by *ar.yw()*, and find the order of the residuals after filtering.

2. Show that *ar.mle()* and *ar.yw()* provide similar analysis for the residuals from the cubic Vostok model, both in the choice of model and in which models are considered close to the best models based on AIC values.

3. In the text, the time stamps for the temperature data were explored to determine the degree of equal or unequal spacing. Compute a "lag" variable for the CO_2 time stamp and perform a similar analysis. How does this compare to the "lag" for the temperature time stamp. Explain why this result is not surprising.

4. Show that, if a model is AR(1) with constant variance and gaps appear in equally spaced data, the ratio of the largest variance to the smallest variance is at most $1/[1 - a]$.

5. A students argues: "It doesn't matter where the filter (A) came from, if the result is quasi-independent observations [observations not distinguishable from white noise based on *ar.yw()* or *ar.mle()*], then the assumptions of the regression model are met." Comment on this claim.

6. In the book the phrase "residuals not distinguishable from white noise" is used a lot. Why would it be less correct to say "residuals that are white noise."

7. Develop a simulation similar to the one in Section 20.5 with $0.65 < R^2 < 0.85$. Use the ARMA(m, l) structure below and pick your own parameters for the periodic functions.

   (i) Simulate the data with $n = 3200$.

(ii) Perform a hypothesis test for the slope coefficient, based on a filtered model.

(iii) Form a random sample of 283 observations from the original 3200.

(iv) Perform a hypothesis test for the slope coefficient, based on a filtered model.

Your class results can be pooled to get an overall assessment of the reliability of filtering in this context.

ARMA(m,l) structures:

(a) $(1 - 0.7B)(1 + 0.1B)\,\varepsilon_n = (1 - [0.4 + 0.4i]\,B)(1 - [0.4 - 0.4i]\,B)w_n$

(b) $(1 - [0.7 + 0.1i]B)(1 - [0.7 - 0.1]B)\,\varepsilon_n = (1 - 0.6B)(1 - 0.3B)w_n$

APPENDIX A

USING DATAMARKET

A.1 OVERVIEW

DataMarket provides a library and package that they have created for exporting data into the R environment. In order to use this package with the library, the user needs to put the following commands into R at the beginning of a session:

```
install.packages("rdatamarket")
library(rdatamarket)
```

The above code will download the "rdatamarket" package and library. It requires about 3.6 megabytes. When the user runs the above code in the R environment, the prompt as shown in Figure A.1 will appear. If the user presses cancel (after the commands cited above), the R code will try to perform the task without the latest updates.

A full list of the commands and functions that the "rdatamarket" package adds can be found at http://cran.r-project.org/web/packages/rdatamarket/rdatamarket.pdf

This code is necessary if the user desires to export data from DataMarket to R.

CRAN stands for Comprehensive R Archive Network. A CRAN mirror a site that contains the latest updates, versions, etc., of R.

If the user desires to select a CRAN mirror, it is advised that a nearby location be selected. For additional information regarding CRAN the user should

Basic Data Analysis for Time Series with R, First Edition. DeWayne R. Derryberry.
© 2014 John Wiley & Sons, Inc. Published 2014 by John Wiley & Sons, Inc.

FIGURE A.1   The CRAN prompt.

visit http://cran.r-project.org/doc/FAQ/R-FAQ.html#What-is-CRAN_003f. This can be found in the Frequently Asked Questions tab in the R-project website which can be found at http://www.r-project.org/.

Once the user has installed or reinstated the "rdatamarket" package at the beginning of the R session, the next step is to have the user create a DataMarket user account. In order to register for an account the user needs to access the "DataMarket" homepage (Figure A.2) at http://datamarket.com/. By pressing the button in the upper left-hand

FIGURE A.2   DataMarket homepage.

corner that is labeled "Log in or sign up (for free)" the user will be taken to a login page (Figure A.3).

At the login page (Figure A.3) a text link exists that will transport the user to the signup page, prompting them for their basic information in order to create an account. After the user's account is created, the user can take full advantage of DataMarket. Below is an example of how to find and transport data from the DataMarket website to the R console.

DataMarket works like a search engine, except that it only brings up data sets. So if a user searches for "New York Temperatures" through the search bar, which can be found on the homepage after an account has been created, DataMarket presents several data sets that contain the words "New," "York," and "Temperature" somewhere in the text of the data. Clicking on the text link titled "Monthly *New York* City

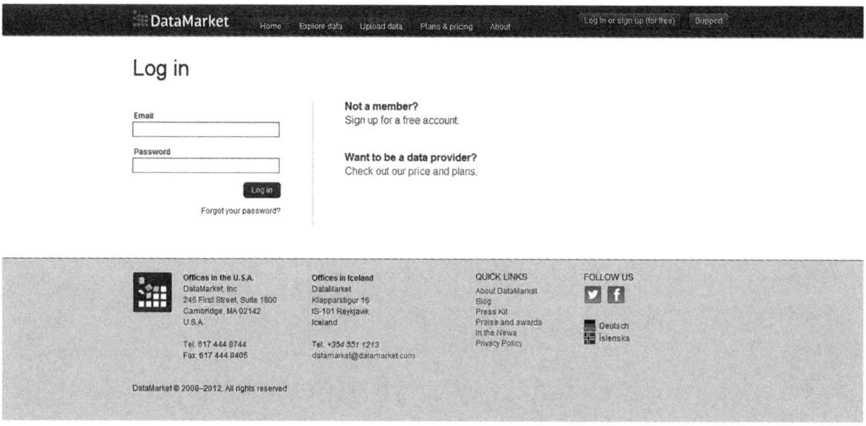

FIGURE A.3   Login page of DataMarket.

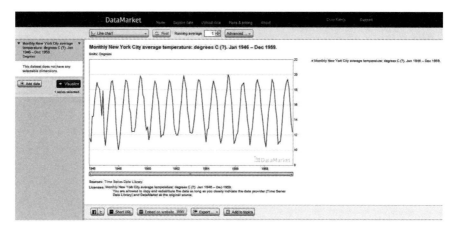

FIGURE A.4   DataMarket's display of New York city temperature data.

average *temperature*: degrees C (?). Jan 1946 – Dec 1959" should transport the user to the following page (Figure A.4).

On this page the user can use export command to send information to either a note pad or the R console itself. By pressing on the Export… button a list box will appear (Figure A.5) above the Export button with several ways in which the data can be exported.

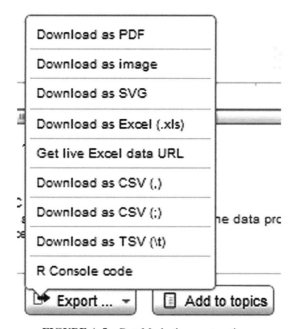

FIGURE A.5   DataMarket's export options.

A.2   LOADING A TIME SERIES IN DATAMARKET

At this point the user can use one of the two options to export the data in the R console: the function *dmlist( )* and the function *dmseries( )*. Each function will be explained, then similarities and differences are discussed.

One of the options in the list box is "R Console Code" (Figure A.6). By clicking on the "R Console Code" list option, the user will be prompted with a message box that contains the code necessary to export, or download, the information into the R console.

There are actually four lines of code here, and if the user were to copy and paste the code directly into the R console for the New York Temperature data set, no harm would be done, although a message may appear *"Warning: package 'rdatamarket' is in use and will not be installed."* The first two lines are, of course, the code already inserted at the beginning of the R session and the third line is an API key. Because this is a secret, it has been blocked out by the author.

If the first two commands have already been inserted, the only code required to load the current data is a <- dmlist("22s0"), which can be cut and pasted into R directly. When using many DataMarket files in a session, it seems more efficient to run the commands installing "rdatamarket" just once per session, so the first three lines are unnecessary.

Of course, after loading the command, the most natural thing to do is explore the object created with names(a): "Month" "Value". The variable "a\$Month" is a date especially useful for graphing time series, but "a\$Value" is just a set of real numbers. The data has a structure found in Figure A.7.

Using *dmlist( )*, a relatively nice plot of the data can be produced quickly (Figure A.8):

```
plot(dmlist("22s0"))
lines(dmlist("22s0"))
```

FIGURE A.6   DataMarket's message box containing R code.

```
> print(a)
        Month    Value
1     1946-01   11.506
2     1946-02   11.022
3     1946-03   14.405
4     1946-04   14.442
5     1946-05   16.524
6     1946-06   17.918
7     1946-07   18.959
8     1946-08   18.309
9     1946-09   18.160
10    1946-10   16.691
11    1946-11   14.480
12    1946-12   17.862
```

FIGURE A.7   R-Console print of *dmlist( )* of New York temperature data.

Nevertheless, an even better graph can be attained by using a function that handles the variable "Month" more properly as a time-based variable, *dmseries( )*.

plot(dmseries("22s0"), xlab = "Time", ylab="Degrees Celsius")

lines(dmseries("22s0")).
This code produces Figure A.9.
The additional labels were possible in the previous graph, but the handling of the *x*-axis is different than, and superior to, *dmlist( )*. A quick check using b <- dmseries("22s0") and names(b) will verify that the *dmlist( )* is better for numerical manipulation and *dmseries( )* is better for graphing the data.

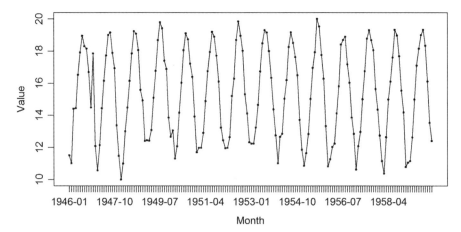

FIGURE A.8   A plot of an the New York temperature data defined by dmlist().

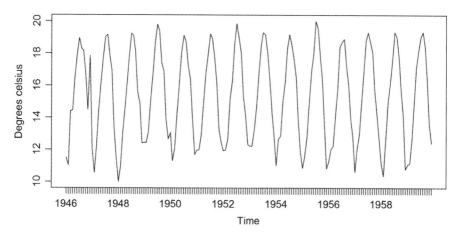

FIGURE A.9   A time plot of the New York temperature data defined by *dmseries( )*.

Finally, the DataMarket library is stored through an online library. The "rdatamarket" package only gives R the ability to fetch information from the online library and put it directly into R. This means that if a user's internet connection is lost, then he or she will not be able to access the online library until the connection is restored.

Of course, it is always easy to access the DataMarket and download the data (Figure A.5) in a number of more traditional methods. For example, it is possible to download the data into a spreadsheet and then copy it into notepad. However, in these formats the date column is of no significance in R.

Detailed information	
Dataset title	Monthly milk production: pounds per cow. Jan 62 – Dec 75
Last updated	26 Jun 2012, 17:36
Provider	Time Series Data Library
Provider source	Cryer (1986)
Source URL	http://datamarket.com/data/list/?q=provider:tsdl
Units	Pounds per cow
Granularity	Month: Jan 1962 – Dec 1975
Language	English
License	**Default open license**
	This data release is licenced as follows: You may copy and redistribute the data. You may make derivative works from the data. You may use the data for commercial purposes. You may not sublicence the data when redistributing it. You may not redistribute the data under a different license. Source attribution on any use of this data: Must refer source.
	View full text →
Description	Agriculture, Source: Cryer (1986), in file: data/milk, Description: Monthly milk production: pounds per cow. Jan 62 – Dec 75

FIGURE A.10   The DataMarket licensing agreement for files in the Times Series Data Library only.

A.3   RESPECTING DATAMARKET LICENSING AGREEMENTS

Each file in DataMarket has licensing conditions associated with that file. Because this book uses only a subset of DataMarket, the Time Series Data Library compiled by Dr. Hyndman of Monash University, the agreement is always the same and quite liberal (Figure A.10). However, two things should be kept in mind, both legally and in terms of common sense: (i) always cite the source when using these files and (ii) remember that other files in DataMarket may have different licenses and different terms of use.

APPENDIX B

AIC IS PRESS!

B.1 INTRODUCTION

The model selection problem in linear regression involves choosing a best model (or at least some best models) from a group of candidate models. One approach being to pick the model that most closely fits the data would appear to involve picking the model with the lowest $\text{SSE} = \sum_i (Y_i - \hat{Y}_i)^2$, but there is an obvious flaw. There is an over-fitting bias in the way SSE is estimated; the criteria used to fit a model (maximum likelihood) minimizes SSE. In particular, if explanatory variables are sequentially added to a regression model, SSE will always shrink, suggesting the absurd conclusion that the most complex model always the one that should be chosen. Put another way, using the same data to both estimate the parameters and subsequently assess fit can produce ever more optimistic estimates of model fit with each additional parameter.

One approach to the model selection problem involves constructively re-stating it as that of picking the model with the best SSE once a correction is made for over-fitting bias. A measure much like SSE that reflects how well the current model might fit fresh data is desired.

B.2 PRESS

Allen (1971, 1974) presents a useful approach to correcting for over-fitting bias in SSE based on a cross-validation idea: $\text{PRESS} = \sum_i (Y_i - \hat{Y}_{(i)})^2$. Computation of

Basic Data Analysis for Time Series with R, First Edition. DeWayne R. Derryberry.
© 2014 John Wiley & Sons, Inc. Published 2014 by John Wiley & Sons, Inc.

PRESS involves splitting the data into a training set of $n - 1$ observations and a validation set of one observation, but does this for all n possible combinations of splits (leave-one-out cross-validation). It is always the case that PRESS > SSE. In general PRESS, as is any reasonable measure motivated by cross-validation, is a better measure of model fit than SSE. Moreover, it is easy to produce data sets in which SSE ≈ 0, but PRESS $\approx$ SST, where SST $= \sum_i (Y_i - \bar{Y})^2$, when the number of explanatory variables approaches the number of observations (so PRESS strongly penalizes extreme over-fitting).

B.3 CONNECTION TO AKAIKE'S RESULT

Akaike's information criterion (AIC) (Akaike, 1974) is also an attempt to correct for over-fitting bias, but in the likelihood (on the $\log_e$ scale). Akaike views this as the problem of using the same data to both estimate the parameters and estimate the resulting log-likelihood. Because the likelihood in a linear regression model has the sufficient statistic SSE, AIC can also be viewed as corrections for over-fitting when (log)-likelihood maximization is used to estimate SSE. Furthermore, Stone (1977) has shown the asymptotic equivalence of leave-one-out cross-validation (the basis for PRESS) and AIC for model selection.

For this reason, it is natural to compute a PRESS-like number based on AIC. In the case of linear regression models: $n \cdot \log_e(\text{PRESS}_{\text{AIC}}/n) = \text{AIC} = n \cdot \log_e(\text{SSE}/n) + 2 \cdot d$, where d is the number of estimated parameters including the variance (Anderson and Burnham, 2002, p. 12) and n is the sample size.

The expression on the right-hand side is the maximized log-likelihood with an adjustment for over-fitting bias, while the left-hand side of the expression is the estimated log-likelihood free of over-fitting bias.

This implies $\text{PRESS}_{\text{AIC}} = \text{SSE} \cdot \exp(2 \cdot d/n)$. It is easy to see that $\text{PRESS}_{\text{AIC}}$ inflates SSE in a way that increases with number of parameters estimated (potentially penalizing over-fitting), but decreases with sample size.

B.4 NORMALIZATION AND R^2

The quantities SSE, PRESS, and $\text{PRESS}_{\text{AIC}}$ all grow with sample size. In model selection normalized values that remain stable over sample size and have simple interpretations are more useful. The explained variation, $R^2 = 1 - \text{SSE}/\text{SST}$, which compares the model variation to the total variation, is a natural way of comparing models, but does share the flaws inherent in SSE. Replacing SSE with either PRESS or $\text{PRESS}_{\text{AIC}}$ provides easy to understand and interpret "bias-corrected value" for R^2. The quantity $R^2_{\text{pred}} = 1 - \text{PRESS}/\text{SST}$ is already standard in regression textbooks and some software. The quantity $R^2_{\text{AIC}} = 1 - \text{PRESS}_{\text{AIC}}/\text{SST}$ has the same interpretation.

Furthermore, when all the observations have equal influence, PRESS = SSE $(n/[n - d - 1])^2$. In this case the expression $R^2_{\text{AIC}} - R^2_{\text{pred}} = (1 - R^2)(2/n + O[1/n^2])$

TABLE B.1   **Four Candidate Regression Models and Three Different Variations of Explained Variation**

Model	R^2	R^2_{pred}	R^2_{AIC}
x_1 only	20.4%	5.3%	6.0%
x_2 only	59.5%	52.4%	52.1%
x_1 and x_2	79.9%	73.8%	74.2%
x_1, x_2 and interaction	79.9%	71.0%	72.0%

for fixed d. In other words, for large sample size and observations with about equal influence, $R^2_{pred} \approx R^2_{AIC}$.

B.5   AN EXAMPLE

The similarity of R^2_{pred} and R^2_{AIC}, even for moderate sample size, is evidenced by numerous examples. Case 9.1 from the Statistical Sleuth (Ramsey and Schafer, 2002) involves just 24 observations with two explanatory variables. One explanatory variable (call it x_1) is coded one-two and the other (call it x_2) is continuous.

The model selection problem has been re-formulated as that of picking the model with the largest "corrected" explained variation (Table B.1). For AIC, this is not different than selecting the model with the lowest AIC value, but adds further meaning to a specific value of AIC. Both the AIC and PRESS calculations show that both the interaction and non-interaction models with both variables are quite close with regard to fit and far superior to the other models. However, the non-interaction model is, strictly speaking, best.

Both PRESS and AIC values perform similarly and very different from SSE. The fitted model, free of over-fitting bias, would account for about 74% (Table B.1) of the total variation in the response values.

B.6   CONCLUSION AND FURTHER COMMENTS

Given the claims made by Stone (1977) and the further calculations presented here, PRESS and AIC should be viewed as almost identical in motivation and results. This is surprising because the detailed derivations appear quite different, with PRESS derived from cross-validation (Allen, 1971, 1974) and AIC derived using Kullback–Leibler distance (Akaike, 1974; Burnham and Anderson, 2002). It would be nearly correct to say they are two radically different derivations of the same result.

Of course, they are not quite identical. PRESS is based on the residuals and leverage values, while PRESS$_{AIC}$ is based on the sufficient statistic alone. When PRESS and PRESS$_{AIC}$ produce dramatically different results, it may be evidence of outliers, influential/leverage points, and/or gross violations of parametric assumptions.

APPENDIX C

A 15-MINUTE TUTORIAL ON NONLINEAR OPTIMIZATION

C.1 INTRODUCTION

It is unlikely that the reader will ever write a program that solves a nonlinear optimization problem, but functions in R (and many other languages) exist for solving
such problems and such functions, in R or otherwise, have many special issues. In
particular, when a function fails to report a useful answer, the user needs to know if the
problem is a bug (poorly written code), or a failure of the called function to converge
to a reliable answer. The purpose of this section is to understand the difficulties in
solving a nonlinear optimization problem, what error messages are likely to occur,
and how to address these errors.

C.2 NEWTON'S METHOD FOR ONE-DIMENSIONAL NONLINEAR OPTIMIZATION

Consider the following problem: min $e^x + x^2$ over all possible values of x. The obvious
option is to take the derivative and set it equal to zero, $f'(x) = e^x + 2x = 0$, but this
does not have a closed form solution. When an optimization problem produces a
derivative or set of first derivatives that cannot be solved as a system of k linear
equations in k unknowns, it is a nonlinear optimization problem. Such problems
usually require iterative solutions.

A method for solving such problems that is often learned in first-year calculus
is Newton's method. The method uses the last guess to produce a next guess and

Basic Data Analysis for Time Series with R, First Edition. DeWayne R. Derryberry.
© 2014 John Wiley & Sons, Inc. Published 2014 by John Wiley & Sons, Inc.

TABLE C.1 Successive Iterations Using Newton's Method

Iteration	x_n	$f'(x_n)$	$f''(x_n)$	x_{n+1}
1	1	4.718	4.718	0
2	0	1	3	−0.333
3	−0.333	0.051	2.717	−0.352
4	−0.352	−0.0007	2.723	−0.352

stops when the guesses no longer change (very much). Each guess is closer to the final answer. In this context, Newton's method is stated as $x_{n+1} = x_n - f'(x_n)/f''(x_n)$. Noting that $f''(x) = e^x + 2$, and using $x = 1$ as an initial guess (not a particularly good guess). The formula is $x_{n+1} = x_n - (e^x + 2x)/(e^x + 2)$. The successive iterations are displayed in Table C.1.

A couple things are worth noting. The iterative process will never produce a first derivative of exactly zero, so judgment is required to decide when to stop, and inspection of the second derivative confirms the solution is indeed a minimum.

C.3 A SEQUENCE OF DIRECTIONS, STEP SIZES, AND A STOPPING RULE

The method is based on local quadratic approximations of the true function. Unless at least partial quadratic approximation is used, the convergence will be slow. At minimum, any method like Newton's method computes a direction and a step size at each iteration. A rule for when to stop in also required.

C.4 WHAT COULD GO WRONG?

Everything!: (1) The formula could have a zero in the denominator, preventing a direction from being found at a given iteration; (2) the local quadratic approximation could be poor, meaning the step will be too far (overshooting the optimum), or nor far enough (taking many iterations); (3) the rule for stopping may be too liberal, stopping before a minimum is reached, or too conservative, taking many iterations to at the final solution, after having found the optimal solution up to the precision of the computer; (4) furthermore, it may be very expensive (computationally) to take the second derivative of the function and evaluate it at every iteration; (5) the procedure may find a local maximum when the user is looking for a local minimum; (6) the starting point for the method could be bad, requiring too many iterations to get to the right solution; (7) finally, due to round off error, it is possible for the direction (after rounding) to be such that the function does not decrease. Perhaps someone can think of others! All of these problems become more of an issue when multiple variables are involved.

While (5) can be a real issue, it is rarely an issue in curve fitting and statistical settings. Usually problems involving minimizing least squares or maximizing likelihood have one global optima of the type the user desires, although there are exceptions which we will not discuss [remember, this is a 15-minute tutorial].

C.5   GENERALIZING THE OPTIMIZATION PROBLEM

Consider a more general problem: minimize $f(x, y) = x^2 + 2xy + y^3$. Ideas like Newton's method still apply, but the quadratic approximation must be made more explicit.

The first derivatives are $\frac{\partial f}{\partial x} = 2x + 2y + 0$ and $\frac{\partial f}{\partial y} = 0 + 2x + 3y^2$.

The second derivatives are $\frac{\partial^2 f}{\partial x^2} = \frac{\partial^2 f}{\partial x \partial y} = \frac{\partial^2 f}{\partial y \partial x} = 2$, $\frac{\partial^2 f}{\partial y^2} = 6y$.

At any point (x_n, y_n), the estimated value of $f\left(x_n + \Delta x, y_n + \Delta y\right)$ is

$$\approx f(x_n, y_n) + \left(2x_n + 2y_n, 2x_n + 3y_n^2\right)\begin{pmatrix} \Delta x \\ \Delta y \end{pmatrix} + \frac{1}{2}(\Delta x, \Delta y)\begin{bmatrix} 2 & 2 \\ 2 & 3y_n \end{bmatrix}\begin{pmatrix} \Delta x \\ \Delta y \end{pmatrix}.$$

Newton's method generalizes to the following:

$$\begin{pmatrix} x_{n+1} \\ y_{n+1} \end{pmatrix} = \begin{pmatrix} x_n \\ y_n \end{pmatrix} - \alpha \begin{pmatrix} \Delta x \\ \Delta y \end{pmatrix} = \begin{pmatrix} x_n \\ y_n \end{pmatrix} - \alpha \begin{pmatrix} 2x_n + 2y_n \\ 2x_n + 3y_n^2 \end{pmatrix}\begin{bmatrix} 2 & 2 \\ 2 & 3y_n \end{bmatrix}^{-1},$$

where $\alpha = 1$. The inclusion of a value α emphasizes that there is a step size that could be regulated as part of the optimization algorithm.

More generally, Newton's method with a variable step size is

$$x_{n+1} = x_n - \alpha \cdot g^T H^{-1},$$

where x_n are vectors, g is a vector of first derivatives, and H is a matrix of second derivatives.

From the point of view of optimization, nothing has changed. The algorithm stops when g is essentially a vector of zeros, and whether the matrix H is positive definite (local minimum), negative definite (local maximum), or indefinite (unclear result) determines the nature of the solution.

C.6   WHAT COULD GO WRONG—REVISITED

Termination: If it is hard to know when a derivative is close to zero, it is harder to know when a vector of derivatives is close to zero. If the requirements for termination are too liberal, the algorithm may stop short of the solution; if the termination requirements are too conservative, the algorithm may continue to fine tune the solution long after the solution has been found to the precision of the computer. As will be seen using $nls()$, it is also possible to terminate based on how slowly the actual function is changing. In other words, if $\left|f(x_{n+1}) - f(x_n)\right| < \varepsilon$, the algorithm terminates.

Bad starting values: almost all nonlinear optimization procedures depend on reasonable initial guesses for the unknown values. If the initial guess is poor, the procedure may give up, even when it is making good progress toward a solution. Fortunately, in most statistical settings, initial guesses are usually easy to come by (see below).

Inversion of H. What allows the Newton's method procedure to find a minimum is that the matrix H is always positive definite (in statistical problem the matrix is related to the covariance matrix and should always be positive definite for least squares minimization and negative definite for maximum likelihood). In reality the matrix may not be invertible; it may fail to be positive definite due to round off error, and/or it may be too expensive to re-compute at every iteration.

There are a number of methods that capture the essence of Newton's method without re-computing H at every iteration. The method of conjugate gradients and quasi-Newton's methods are two examples.

Although H is, in theory invertible, for most statistical problems, computers can have round off error. Many problems related to matrix inversion, and matrix round off error, in general, are indicated when the condition number of the matrix is large. The R function *kappa()* gives the approximate condition number of a square matrix.

According to Numerical Mathematics and Computing (Cheney and Kincaid, 2008): "As a general rule of thumb, if the condition number $kappa(X) = 10^k$ [replacing their expression with R code] then you may lose up to k digits of accuracy on top of what would be lost to the numerical method due to loss of precision from arithmetic methods."

Poor quadratic approximation. The method, in all its forms, is based on local quadratic approximations. Some functions are poorly represented by quadratic approximations in at least some regions where the function exists. If the local approximation is poor, the initial guess as to the direction and step size will be poor. This will result in a procedure taking too many iterations.

C.7   WHAT CAN BE DONE?

Of course, most of the time none of these bad things occurs, however when they do occur, the user needs to know which of these bad things has happened and what remedies are available. Most bad messages are about singular matrices, failure to converge, or stopping based on iteration limits.

Now it is possible to look at the nuts and bolts of a specific function and understand what kind of bad output could occur.

Under *help(nls)* the following can be found:

trace   logical value indicating if a trace of the iteration progress should be printed.
          Default is FALSE. If TRUE the residual (weighted) sum-of-squares and the
          parameter values are printed at the conclusion of each iteration. When the
          "plinear" algorithm is used, the conditional estimates of the linear parameters
          are printed after the nonlinear parameters. When the "port" algorithm is used
          the objective function value printed is half the residual (weighted)
          sum-of-squares.

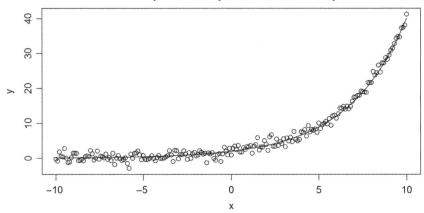

FIGURE C.1   Simulated exponential data.

If an application is going badly, it is possible to follow each step of the procedure by including trace = TRUE to the function call (this feature exists because people need to use it!).

Consider the following simulated data for a highly nonlinear relationship, displayed in Figure C.1:

```
a <- 2
c <- 0.3
x <- c(-100:100)/10
error <- rnorm(length(x),0,1)
y <- a*exp(c*x)+error
y_fit <- a*exp(c*x)
plot(x,y)
lines(x,y_fit)
title("Example of an exponential relationship")
```

The function *nls()* could be used to find a best fitting line through the points. Solving the optimization problem with *nls()*, with bad starting values, fit <- nls(y ~ a*exp(c*x), trace=TRUE, start=list(a=10,c=-2)), yields the following output:

```
7.139811e+19 :  10 -2
87107.93 :   2.881669e-07 -2.000000e+00
28293.54 :   2.966834e-08 -1.989456e+00
27847.54 :   3.335262e-08 -1.874325e+00
27779.37 :   1.178617e-07 -1.512263e+00
27770.41 : 2.745499e-06 8.303805e-01
Error in numericDeriv(form[[3L]], names(ind), env) :
Missing value or an infinity produced when evaluating the model
```

It is easy to check [*help(numericDeriv)*] that a function called by *nls()*, *numericDeriv()* cannot compute the derivatives at this iteration.

Reasonable starting values are not hard to find. Although the normality condition of the errors is violated, a least squares approach can be applied:

fit_ols <- lm(log(y+2)~ x)   # 2 added to y to produce all positive values

	Estimate	Std. Error	t value	Pr(>|t|)
(Intercept)	1.678255	0.028037	59.86	<2e-16
x	0.154784	0.004832	32.03	<2e-16

These estimates are quite crude, but good enough to get the job done:

fit <- nls(y ~ a*exp(c*x), trace=TRUE, start=list(a=5.35,c=0.155))
the guess for *a* is exp(1.678).
3068.85 : 5.350 0.155
1384.038 : 3.5666708 0.2104668
840.1805 : 1.7933324 0.2928588
189.4892 : 2.0158136 0.2997164
187.798 : 2.0155112 0.2988247
187.798 : 2.0157555 0.2988066

Given the true parameter values are 2.0 and 0.3, the results here are quite good. It is easy to check that the quantity 187.798 is the sum of the squared residuals, and it would seem the algorithm terminates when this is no longer changing.

Within *help(nls)* is a further topic *nls.control()*. This should be easy to understand now:

Allow the user to set some characteristics of the nls nonlinear least squares algorithm.

Usage

nls.control(maxiter = 50, tol = 1e-05, minFactor = 1/1024,
    printEval = FALSE, warnOnly = FALSE)

Arguments

maxiter	A positive integer specifying the maximum number of iterations allowed.
tol	A positive numeric value specifying the tolerance level for the relative offset convergence criterion.
minFactor	A positive numeric value specifying the minimum step-size factor allowed on any step in the iteration. The increment is calculated with a Gauss-Newton algorithm and successively halved until the residual sum of squares has been decreased or until the step-size factor has been reduced below this limit.
printEval	A logical specifying whether the number of evaluations (steps in the gradient direction taken each iteration) is printed.

warnOnly	A logical specifying whether nls() should return instead of signalling an error in the case of termination before convergence. **Termination before convergence happens upon completion of maxiter iterations, in the case of a singular gradient, and in the case that the step-size factor is reduced below minFactor.**

Obviously the usual suspects are here. There is a maximum number of iterations, there is a tolerance criterion (in this case when SSE, the sum of the squared residuals, changes less than "tol" from iteration to iteration, it is assumed that no substantial further progress can be achieved). The step size can be halved at each step until a certain limit, "minFactor," is reached. If the quadratic approximation is horrible, it is possible that the initial step sizes are too big, in which case "minFactor" can be decreased. Of course *nls()* calls other functions, and termination can occur due to the other functions as well (as already seen).

If termination is due to the maximum number of iterations being used, it is easy to follow trace and observe if real progress is being made. If so, it is easy to use the last values from the previous trace as initial values for the next attempt.

If singularities occur while running *nls()*, brutal as it may sound, it may be wise to compute the matrix of second derivatives and determine the condition number at the final point in "trace." One way it might help to reduce the condition number is to re-scale or reformulate the problem.

A couple of other issues come to mind. It is possible to improve the condition number of the H matrix using regularization, replace H with $H + \varepsilon I$, where I is the identity matrix and ε is "a small number." This is, for example, the basis for ridge regression. This cannot be done within *nls()*, as far as I know. If the current formulation of the problem is highly nonquadratic, locally, this will cause the step sizes to be very small at each iteration. If one knows that the algorithm is making good progress, reduce minFactor.

A simple re-scaling of the problem can, occasionally, have many benefits. Recall, *nls()* appears to be using the criteria: $|\text{SSE}(x_{n+1}) - \text{SSE}(x_n)| < \text{tol}$. Re-scaling the data is essentially the same as changing tol. It might be better to think of the problem, including scaling, as one of relative error: $|\text{SSE}(x_{n+1}) - \text{SSE}(x_n)|/\text{SSE}(x_{n+1}) < \varepsilon$ so $|\text{SSE}(x_{n+1}) - \text{SSE}(x_n)| < \varepsilon \cdot \text{SSE}(x_{n+1}) = \text{tol}$. That is, tolerances should be chosen to control termination with relative error in mind. In the application above, implicitly $\varepsilon = 10^{-5}/187.798 = 5.3 \cdot 10^{-8}$, and although the algorithm only took a few iterations, this was probably an unnecessary level of precision. It is difficult for the author to imagine many situations where $\varepsilon = 10^{-4}$, in the context of relative error, is not sufficient.

In general, increasing the number of iterations (maxiter), increasing the tolerance (tol), and reducing the possible step size (minFactor), all increase the chances of getting a final answer without any early termination warnings, but they do not increase your chances much of getting a reliable answer, unless one knows why early termination is occurring in the first place.

The student can have a lot more control over the optimization process by writing their own code. This is not always that difficult.

REFERENCES

Aho K, Derryberry D, Peterson T, Model selection for ecologists: the worldview of AIC and BIC. *Ecology* 2014, 95: In press.

Akaike H. A new look at the statistical model identification. *IEEE Transactions on Automatic Control* 1974, 19:716–723.

Allen DM. Mean square error of prediction as a criterion for selecting variables. *Technometrics* 1971, 13:469–475.

Allen DM. The relationship between variable selection and data augmentation and a method for prediction. *Technometrics* 1974, 16:125–127.

Annenberg Learner, Against All Odds: Inside Statistics, Unit 14: The Question of Causation. Available at: http://www.learner.org/courses/againstallodds/unitpages/index.html (accessed November 7, 2013).

Baldi B, Moore DS. *The Practice of Statistics in the Life Sciences*. W H Freeman, 2012.

Bloomfield P. *Fourier Analysis of Time Series*, 2nd edn. New York: John Wiley & Sons, Inc., 2000.

Box G, Jenkins G, Reinsel G. *Time Series Analysis*, 4th edn. New York: John Wiley & Sons, Inc., 2008.

Broemeling LD, Studies in the History of Probability and Statistics: Semmelweis and Childbed Fever. Available at: http://www.mdanderson.org/education-and-research/departments-programs-and-labs/departments-and-divisions/division-of-quantitative-sciences/research/biostats-utmdabtr00504.pdf (accessed September 15, 2013).

Burnham K, Anderson D. *Model Selection and Multimodel Inference*, 2nd edn. New York: Springer, 2002.

Cheney W, Kincaid D. *Numerical Mathematics and Computing*, 6th edn. Brooks/Cole, 2008.

Devore J, *Probability and Statistics for Engineering and the Sciences*, 7th edn. Thomson Higher Education, 2008.

Edwards AWF. *Likelihood*. Cambridge University Press, 1972.

Kass RE, Raftery A. Bayes factors. *Journal of the American Statistical Association* 1995, 90(430):773–795.

Kitagawa G. *Introduction to Time Series Modeling*. New York: CRC Press, 2010.

Luenberger D. *Linear and Nonlinear Programming*. Addison-Wesley, 1984.

Mandal R, St-Hilaire S, Kie JG, Derryberry D, Spatial trends of breast and prostate cancers in the United States between 2000 and 2005. *International Journal of Health Geographics 2009,* DOI:10.1186/1476-072X-8-53

Nash S, Sofer A. *Linear and Nonlinear Programming*. McGraw-Hill, 1996.

Petit JR., Jouzel J, Raynaud D, Barkov NI, Barnola JM, Basile I, Bender M, Chappellaz J, Davis J, Delaygue G, Delmotte M, Kotlyakov VM, Legrand M, Lipenkov V, Lorius C, Pépin L, Ritz C, Saltzman E, Stievenard M. Climate and atmospheric history of the past 420,000 years from the Vostok Ice Core, Antarctica. *Nature* 1999, 399:429–436.

Picard RR, Berk KN, Data splitting. *The American Statistician* 1990, 44:140–147.

Picard, RR, Cook, RD, Cross-validation of regression models. *Journal of the American Statistical Association* 1984, 79:575–583.

Ramsey FL, Schafer DW. *The Statistical Sleuth*, 2nd edn. Pacific Grove: Duxbury, 2002.

Royall R, *Statistical Evidence: A Likelihood Paradigm*. London: Chapman & Hall, 1997.

Silver N. *The Signal and the Noise*, New York: Penguin Press, 2012.

St-Hilaire S, Mandal R, Commendador A, Mannel S, Derryberry D. Estrogen receptor positive breast cancers and their association with environmental factors. *International Journal of Health Geographics* 2011. DOI:10.1186/1476-072X-10-32.

St-Hilaire S, Mannel S, Commendador A, Mandal R, Derryberry D. Correlations between meteorological parameters and prostate cancer. *International Journal of Health Geographics* 2010. DOI:10.1186/1476-072X-9-19.

Stone M, An asymptotic equivalence of choice of model by cross-validation and Akaike's criterion. *Journal of the Royal Statistical Society B* 1977, 39:44–47.

Wikipedia, Wolf Sunspot Number. Available at: http://en.wikipedia.org/wiki/Wolf_number (accessed August 21, 2013).

Wikipedia, Semmelweis. Available at: http://en.wikipedia.org/wiki/Ignaz_Semmelweis (accessed September 10, 2013).

Wikipedia, Solar Storm of 1859. Available at: http://en.wikipedia.org/wiki/Solar_storm_of_1859 (accessed September 11, 2013).

Wikipedia, Little Ice Age. Available at: http://en.wikipedia.org/wiki/Little_Ice_Age (accessed September 15, 2013).

Yang Y, Can the strengths of AIC and BIC be shared? A conflict between model identification and regression estimation. *Biometrika* 2005, 92(4):937–950.

INDEX

Note: Locators followed by "*f*" and "*t*" refer to figures and tables respectively.

acf( ), 6, 39*f*
    autocovariance estimation coding,
        38–39
    background, 37–38
    and spectrum, 68–69
    for white noise errors, 68*f*
acos( ), 6
AIC. *See* Akaike's information
        criteria
Akaike's information criteria (AIC),
        106–107, 282
    as cross-validation, NYC temperatures,
        112
    model selection with, 117–118
anova( ), 10
arima.sim( ), 6, 130, 159–160
ARMA(2,2) model, 166–167
AR(*m*) filtering matrix, 187–189
    filtering information, 189
    linear algebra, 187
    and lm( ), 188
    to model MA(3), 181–184
    standard computations, 188

AR(1) model for irregular spacing
    final analysis, 268–269
    method, 266
    motivation, 265–266
    results, 266–267
    sensitivity analysis, 267
AR(*m*) structure, residuals for
    data display, 235–236, 235*f*–236*f*
    filtering twice, 236–238
ar.yw( ), 6, 171–174
asin( ), 6
Assumptions
    equal variance
        regression, 31
        two- sample *t*-test, 31
    independence, 31
    introduction, 28–29
    logarithmic transformations, illustration
        of, 32–34
    normality, 29–30
        heavy tails, 30–31
        left skew, 30
        right skewed, 30

Basic Data Analysis for Time Series with R, First Edition. DeWayne R. Derryberry.
© 2014 John Wiley & Sons, Inc. Published 2014 by John Wiley & Sons, Inc.

atan( ), 6
Autocorrelation
    AR(1), 41
    AR(2), 41–42
    estimation, 37
    for MA(1) models, 51–52
    for MA(2) models, 52
    stationarity, 36
Autocovariance
    AR(1), 41
    AR(2), 41
    ARMA*(m,l)* model, 169–170
    estimation, 37, 38
    properties, 36–37
    stationarity, 36
    white noise, 37
Autoregressive model of order 1, AR(1),
        35
    adjustments, 125–151
        implications, 151–152
        skip method, 152
    autocorrelation, 41
    autocovariance, 41
    definition, 40
    examples (stable and unstable models),
        44–46, 45*f*–46*f*
    illustration, 42
Autoregressive model of order 2, AR(2),
        35
    autocorrelation, 41–42
    autocovariance, 41
    examples, 46–50, 46*t,* 47*f*–49*f*
    and power spectrum, 68–70, 70–72
    preliminary facts, 40
    R code, 42–43
    simulating data, 42–44

Backshift operator
    and ARMA*(m,l)* models, 162
    definition, 161
    examples, 162–164
    stationary condition for AR(1) model,
        161
Bayesian information criteria (BIC),
        106–107
Best linear unbiased estimators (BLUES), 9
BIC. *See* Schwarz information criteria
BLUES. *See* Best linear unbiased
        estimators

Boise river flow data, 121*f*–122*f*
    data splitting, 123
    model selection with AIC, 122–123,
        123*f*
    model selection with filtering, 147–151
    residuals, 123–124, 124*f*
Breast cancer, data analysis
    background, 228–229
    estrogen response negative, 228–229,
        229*t*
    estrogen response positive, 228–229,
        229*t*
    and female colon cancer, 241*f*
    first data set (1992–2001), 229–232
    second data set (1975–2005)
        background, 232–233
        data structure, 232*t*
        data trend, 233–235, 234*f*–235*f*
        regression analysis with filtered data,
            238–243
        residuals for AR*(m)* structure,
            235–238
        statistical analysis, 233

Carrington, Richard, 214
Complex conjugates, 62
Complex numbers, 62
    magnitude of, 62–63
Complex periodic model
    accidental deaths
        data splitting, 119–120
        Fourier series structure, 116
        model selection with AIC, 117–118
        model selection with likelihood ratio
            tests, 118–119
        periodic data, comments on,
            120–121
        R Code, fitting large Fourier series,
            116–117
        residual, 120*f*
        training set model, 120*f*
        validation set model, 120*f*
    monthly river flows, furnas 1931–1978
        AR*(m)* filtering matrix, 187–189
        data, 185
        data splitting, 191–192
        model selection, 189, 189*t,* 192–193,
            192*t,* 193*f*
        periodic model, 123*f*

predictions for AR*(m)*, 190
saturated model, 186–187, 187*f*
Comprehensive R Archive Network
(CRAN), 283
Coronal mass ejections, 213
cos( ), 6, 66, 90
CRAN. *See* Comprehensive R Archive
Network
Creek, Gregory, 248
Crosby, Ben, 245
Cross-validation, NYC temperatures
AIC for, 112
data splitting, 108–110, 109*t*, 110*f*
explained variation, R_2, 108
leave-one-out cross-validation,
110–112

Data import, 7
DataMarket, 21
export options, 276
homepage, 275*f*
licensing agreements, 280
login page, 275*f*
overview, 273–276
time series loading, 277–279
Data simulations, 9
Data splitting, 119–120, 191–192
d^c, 6
45-Degree line model, 104–106
dmlist(), 277
New York temperature data plot,
288*f*
dmseries( ), 277
New York temperature data, 279

Endocrine disruptors, 228
Equal variance assumption
regression, 31
two- sample *t*-test, 31
ER+. *See* Estrogen response positive
ER-. *See* Estrogen response negative
Estrogen response negative (ER-), 228
breast cancer, 228–230
first data set (1992–2001), 229–231
rates, 230–231
second data set (1975–2005), 232–243
Estrogen response positive (ER+), 228
breast cancer, 228–230
first data set (1992–2001), 229–231

rates, 230–231
second data set (1975–2005), 232–243
Euler's formula, 63
exp( ), 6
Explained variation, R_2, 108
Export options, 276

Fast Fourier transform (FFT), 75–77
Female colon cancer, 241*f*
FFT. *See* Fast Fourier transform
Filtering, 125, 133–134
and Boise river flow data, 147–151
comments on, 137–138
and global warming model, 136,
136*f*
floor( ), 85
"For" statement, 5–6
Fourier series, 116–118, 119*f*
Fourier series structure, 116
Functions (R)
acos( ), 6
asin( ), 6
atan( ), 6
cos( ), 6
d^c, 6
exp( ), 6
log( ), 6
pi, 6
sin( ), 6
sqrt( ), 6
tan( ), 6
see also Time series, functions

General ARMA models
arima.sim( ), 159–160
and backshift operator, 162
examples, 160–161, 160*f*
mathematical formulation, 159
representative collection, 160*f*
spectrum for, 175–177
Geometric series, 63

Hat matrix, 111
Heavy tails, 30–31
help( ), 10
help(numericDeriv), 289
High elevation (snow), 245
Homepage, 275*f*
Hyndman, Rob, 24

"If" statement, 5
Impulse response operator
    computation
        coefficients computation, 165–166
        definition, 165
        plotting, 166–167
    interpretation, 167
    intuition, 164
    utility, 167
Influential points, 13
Information criteria
    Akaike's information criteria, 106–108
    and model selection, 173–174, 173t
    Schwarz information criteria,
        106–108
Inquiry functions
    anova( ), 10–12
    help( ), 10–12
    names( ), 10–12
    summary( ), 10–12
International sunspot number, 213
Intervention model
    directory assistance
        concern, 199
        data, 199
        filtering information, 199–200,
            202
        model selection, 200, 200t
        saturated model, 199–200
    ozone levels in Los Angeles, 202–205,
        203f, 204f, 204t
    structure, 198

kappa( ), 287

Leave-one-out cross-validation, 110–112
Left skew, 30
Leverage points, 13
Licensing agreements, 280
Likelihood ratio tests, 101–104
    model selection with, 118–119
Linear model, 102–104, 102f
lm( ), 9–10, 214
    vs. nls( ), 216
log( ), 6
Login page, 275f
lowess( ) function, 81–82
Low (rain) elevation watersheds, 245
    initial fits for, 247f

Matrix manipulation, in R
    commands, 16
    OLS, 15–16
mean(x), 4
Mid (mixed) elevation watersheds,
        245–248
    initial fits, 247f
Modeling
    algorithm, 180
    assumption, 180–181
    example
        AR(m) filter to model MA(3),
            181–184
        CO_2 levels at Mauna Lau, 193–198
        monthly river flow, 185–193
    skip method, 184–185, 184t
Model selection
    with AIC, 117–118
    with likelihood ratio tests, 118–119
Monthly river flow, complex periodic
        model
    AR(m) filtering matrix
        filtering information, 189
        fitting a model with lm( ), 188
        linear algebra, 187
        standard computations, 188
    data, 185
    data splitting
        computations, 191–192
        linear algebra, 191
        overview, 191
    model selection, 189, 192–193
    predictions for AR(m) model, 190
    saturated model, 186–187
Moving average model, MA(1)
    acf( ) plots, 55, 55f–56f
    and AR(m) models, 52
    autocorrelation for, 51–52
    simulated examples, 52–54, 53f–54f
Moving average model, MA(2)
    acf( ) plots, 54–55, 55f
    autocorrelation for, 52
    simulated examples, 54

Naïve analysis
    CO_2 and temperature change
        association, 258–259
    model selection, 258
    saturated model, 256–257

Naïve code, 72–74
names( ), 10–11
Naming conventions, 4
Nested models, 99–101
Newton's method (for nonlinear
        optimization), 284–285
nls( ), 214–216, 289
    vs. lm( ), 216
Noise, 18
Nonlinear optimization, tutorial on
    general problem, 286
    introduction, 284
    Newton's method for, 284–285
    revisit, 286
Normality assumption, 29
    heavy tails, 30–31
    left skew, 30
    right skew, 30
numericDeriv( ), 289
NYC temperatures
    application, 146–147
    AR(1) prediction model, 144
    cross-validation
        Akaike's information criterion, 112
        data splitting, 108–110
        explained variation, 108
        leave-one-out cross-validation,
            110–112
    data, 142–144
    outlier, 92, 92*f*
    periodic function fitting, 91, 91*t*
    prediction intervals, 142–144
    simulation, 144–146

Observatory factor, 213
OLS. *See* Ordinary least squares
Ordinary least squares (OLS), 8–9

pacf( ), 6, 174–175
Partial autocorrelation plot
    hypothesis tests sequence, 174, 174*t*
    pacf( ) function, 174–175, 175*f*
Periodic function fitting, 91
Periodic models, 90–91
    complications
        accidental deaths, 96
        CO_2 data, 93–94
        sunspot data, 94–96
    daily average, 205–206, 206*f*

example (NYC temperature data),
        90–91, 91*t*
    outlier, 92
    periodic function fitting, 91
    refitting, 92–93
    monthly average, 206–207, 207*f*
    weekly average, 205–206, 206*f*
Periodic transcendental functions, 64
Periodogram, 65
    and *acf( )* plot, 78*f*
    example, 74–75
    Naïve code for, 72–74
    periodic analysis, 85–86
    periodic behavior, 73*f*
    for power spectrum, 68
    and smoother, 79–80
    and white noise, 73*f*
Personal reduction coefficient *(K),* 213
Phase, 89
Pi, 6
Power spectrum, 65
    and *acf( )* plot, 68*f*
    for ARMA processes, 175–177
    for AR(1) models, 68–70, 69*f*–70*f*
    for AR(2) models, 70–72
    and autocorrelation function, 66–67
    definition, 66
    and periodogram plot, 68*f*
    for white noise, 68
Predictions for AR*(m),* 190
PRESS, 281–282
Prostate cancer, data analysis
    background, 228–229
    estrogen response negative, 228–229
    estrogen response positive, 228–229
    first data set (1992–2001), 229–232
    second data set (1975–2005)
        background, 232–233
        data structure, 232*t*
        data trend, 233–235
        regression analysis with filtered data,
            238–243
        residuals for AR*(m)* structure,
            235–238
        statistical analysis, 233
Prostate-specific antigen (PSA), 234
PSA. *See* Prostate-specific antigen
Pseudo-periodic model, 161
p-values, 29, 100

qqnorm( ), 13, 14*f*, 15
Quadratic model, 101–104, 101*t*
Quasi-independent observations, 137

R (programming language)
    code, 3
    common functions, 6
    console code, 277
    conventions, 5
    data sources, 24–26
    inquiry functions, 10–12
    matrix manipulation, 15–16
    model parameters estimation, 9–12
    smoothers in
        lowess( ), 81–82
        smooth.spline( ), 82–83
    structures, 5
R Code, fitting large Fourier series, 116–117
rdatamarket package, 283
read.csv( ), 7
read.delim( ), 7
read.table( ), 7, 25
Real data, 21
Refitting, 92–93
Regression, 31
Regression model
    matrix representation, 9
    OLS estimates, 9
    ordinary least squares, 8–9
    for periodic data, 89–96
Relative sunspot number, 213
Residuals analysis, 14*f*
    influential points, 13
    lack of fit, 13
    nonwhite noise error, 13–14
    normality, 13
    outliers, 13
    plots, 14–15, 14*f*
    unequal variance, 14
Richer models, 116
Right skew, 30
    p-values, 30
    and unknown period, 95*f*
R^2pred, 111

Saturated model, 117, 193–194
    and data fit, 197–198
    and filter, 193–194

naïve analysis, 256–257
    pruning, 196–197
    residuals, 195*f*
scan( ), 7, 25
Semmelweis, Ignaz Philipp, 138
Semmelweis data, 22–24, 25–26
Semmelweis intervention
    data, 138–139
    filtered analysis, 140–142
    inferences, 142
    serial correlation, 139
    transformations, 142
    vs. patch/uncut case, 139–140
Serial correlation
    and Semmelweis intervention, 139
Signals, 18
Simple mean model, 104–106
Simple regression, 20
    analysis of variance, 100*t*
    hypothesis tests, 99–101
    ratio tests, 101–104
    Sleuth case, global warming
        analysis, 135–136, 135*f*, 136*t*
        data, 132–133
        filtering, 133–134, 137–138
        simulation, 134–135, 135*t*
Simulated data, 20–21
sin( ), 6
Skip method, 152, 184–185
Smoothers, 80
    lowess( ) function, 81–82
    for series
        known period, 85–86, 86*f*
        unknown period, 86–87, 86*f*
    smooth.spline( ) function, 82–83
smooth.spline( ) function, 82–83
Solar flares, 213–214
solve( ), 16
spans( ), 84–85
spec.pgram( ), 6, 75–77
sqrt( ), 6
SSE, 98
SST, 282
Standard errors, 13
Statistical operations, 4
Straight-line model, 104–106
summary( ), 10, 12
sum(x), 4

tan( ), 6
Tennant, Christopher, 245
Time series, 19
    assumptions, 143
    data, 21–24
    extrapolation, 143
    prediction intervals, 143
Time Series Data Library, 21, 24
Time series function (R)
    acf( ), 6
    arima.sim( ), 6
    ar.yw( ), 6
    pacf( ), 6
    spec.pgram( ), 6
    ts( ), 6
Time series loading, 277–279
Transcendental series, 63
ts( ), 6, 25
t-tests, 23
Two-sample *t*-test, 23, 29, 31
    adjustment for AR(1), 128–129
    assumption, 31
    simulation example, 129–130,
        131*f*
    Sleuth data, 125–128
    Sleuth data analysis, 131–132

Variable lag, 263
Vostok ice core data
    alignment
        issues, 254
        matched dates, 254, 255*f*
        need, 254
        patterns, 254, 255*f*
        time stamps, 255–256, 256*f*
    AR(1) model for irregular spacing
        final analysis, 268–269
        method, 266
        motivation, 265–266
        results, 266–267
        sensitivity analysis, 267, 268*t*
    naïve analysis
        CO_2 and temperature change
            association, 258–259
        model selection, 258, 258*t*
        saturated model, 256–257, 257*t*

related simulation
    code, 260
    model, 259–260, 261
    sample of 283, 262–265, 265*t*
source, 251–252, 252*t*

Watersheds data
    averaging data, 248–249
    fitting Fourier series
        data structure, 246
        data to physical processes,
            connecting patterns in, 246–248
        Fourier series fits to data, 246
    high elevation (snow), 245–248
    low (rain) elevation, 245–248
    mid (mixed) elevation, 245–248
    results, 250
White noise, 18, 23, 24
    and autocovariance, 37
    and power spectrum, 68
Wolf, Rudolf, 213
Wolf number, 213
    amplitude, instability in, 217–218
    background, 213–214
    data splitting (for prediction)
        approach, 220–222
        AR-adjusted predictions, 223
        AR correction, 222
        fitting one step ahead, 222, 223*f*, 224*t*
        model selection, 223–224
        predictions two steps ahead,
            224–225, 225*f*, 225*t*
    mean, instability in, 217–218
    nls( ) function, 214–216
    period, instability in, 217–218, 218*f*
    period determination, 216–217
    sunspot data, 217*f*
    for unknown period, 214

Yule–Walker equations
    AR*(m)* and, 170–174
    errors sequence, 172*f*
    model selection (using information
        criteria), 173–174, 173*t*

Zürich number, 213